Power, Profits, and Patriarchy

Power, Profits, and Patriarchy

The Social Organization of Work at a British Metal Trades Firm, 1791–1922

William G. Staples and Clifford L. Staples

ROWMAN & LITTLEFIELD PUBLISHERS, INC.
Lanham • Boulder • New York • Oxford

ROWMAN & LITTLEFIELD PUBLISHERS, INC.

Published in the United States of America
by Rowman & Littlefield Publishers, Inc.
4720 Boston Way, Lanham, Maryland 20706
www.rowmanlittlefield.com

12 Hid's Copse Road, Cumnor Hill, Oxford OX2 9JJ, England

British Library Cataloguing in Publication Information Available

Library of Congress Cataloging-in-Publication Data

Staples, William G.
    Power, profits, and patriarchy : the social organization of work at a British metal
trades firm, 1791–1922 / William G. Staples and Clifford L. Staples.
        p. cm.
    Includes bibliographical references and index.
    ISBN 0-7425-0076-4 (alk. paper) — ISBN 0-7425-1640-7 (pbk. : alk. paper)
        1. Metal trade—Great Britain—History. 2. Metal trade—Management—Case
studies. I. Staples, Clifford L., 1953– II. Title.

HD9506 .G72 S73 2001
338.7'672'094249—dc21

                                                                        2001041701

Printed in the United States of America

⊖™ The paper used in this publication meets the minimum requirements of American
National Standard for Information Sciences—Permanence of Paper for Printed Library
Materials, ANSI/NISO Z39.48-1992.

For Ian and Caitlin

Contents

Preface

This book is about changes in the character of industrial organization and relations at a British metal trades firm. It is about the capacity of both the factory owners and the laboring men and women in their workshops to claim, advance, or defend their respective interests in the production process. And it is about the political, ideological, and rhetorical means by which those investments were articulated, shaped, regulated, guarded, and promoted. By tracing the experience of this one firm over the course of more than a century, our intention is to illuminate the conditions under which these social relations of production appeared more or less conflictual or consensual. More specifically, we focus our greatest attention on the circumstances under which people associated with the firm—both the owners and various configurations of workers—defined shared interests, and acted, or failed to act, together on those interests. By balancing a concern with the materiality of production—the process of transforming raw materials into useful products—with its ideological, cultural, and political moments, this book offers new insights into the nature of work as well as social and class relations.

We have for so long spoken of an "industrial revolution" founded on technological change and so sweeping and powerful that it consumed everything and everyone in its path; it is only quite recently that we have begun to incorporate historical accounts that are sensitive to the diversity of social and economic arrangements that emerged within the context of this "great transformation." As a result, we have come to appreciate more fully the role of human agency and to see this "revolutionary" *moment* as far more of a lurching *process*, drawn-out and filled with alternative routes to what would become the modern "factory." This book chronicles the route taken by one Midlands metal trades firm during a crucial period of British capitalism.

Writing this book has been a bit of a "lurching process" itself. Bill began the book fifteen years ago when he was a graduate student at the University of Southern California. But it languished for several years while other projects were completed, so we decided to finish the book together. We did not, however, do it alone. The project received financial support from both the General Research Fund and the Hall Center for the Humanities at the University of Kansas; a Schmitt Grant for Research in European, African, or Asian History from the American Historical Association; the Senate Scholarly and Creative Activities Committee at the University of North Dakota; and the personal generosity of E. R. Warren. We appreciate the generous assistance provided by the library staffs and archivists at the University of Southern California; University of California, Los Angeles; University of North Dakota; Watson and Spencer Libraries, the University of Kansas; Center for British Studies, University of Colorado, Boulder; British Museum Reading Room; British Library of Political and Economic Sciences; British Library, Newspaper Collection, Colindale; Birmingham Reference Library; Modern Records Centre, University of Warwick; Sandwell Local Studies Centre; and the Black Country Museum, West Bromwich, where one can find the papers and artifacts of Archibald Kenrick and Sons. Research assistance was provided by Su Lee Ho, Nick Thomas, and Janet Rex. Cliff's second trip to England was more fun and productive because of Janet's help and companionship. We are also grateful to Janet for her careful editorial work. Pam Lerow and Lynn Porter of the CLAS Word Processing Center of the University of Kansas did a wonderful job preparing the manuscript for production. And thanks to our editor, Dean Birkenkamp, for supporting the project and the staff of Rowman & Littlefield Publishers for all their good work.

We have adapted some previously published material for this book. Parts of chapters 1 and 2 were published by William G. Staples in "Technology, Control and the Social Organization of Work at a British Hardware Firm, 1791-1891," *American Journal of Sociology* 93, no.1 (1987): 62-88, copyright © 1987 by The University of Chicago. All rights reserved. And chapter 4 draws heavily on Clifford L. Staples and William G. Staples, "'A Strike of Girls': Gender and Class in the British Metal Trades, 1913," *Journal of Historical Sociology* 12, no. 2 (1999): 158-180, copyright © Blackwell Publishers, Ltd. Both are reprinted here with kind permission of the publishers.

Over the years, many colleagues and friends offered their ideas, comments, and criticisms on various parts of this work. We would like to thank Michael Burawoy, Linda Fuller, Barbara Laslett, Jon Miller, Carol Warren, Sonya Rose, R. A. Church, David N. Smith, Bob Antonio, Tom Weiss, Tony Rosenthal, Wendelin Hume, Michael Schwalbe, Kathleen Tiemann, Michael Hanagan, Maurice Zeitlin,

Judy Stepan-Norris, Jonathan Zeitlin, Joey Sprague, Joshua Rosenbloom, Angel Kwolek Folland, Ann Schofield, Sandi Abrecht, Barry Shank, Surendra Bhana, Carl Strikwerda, Victor Bailey, Eric Hanley, James Mochoruk, Barbara Handy-Marchello, Lizette Peter, and Ann Kelsch.

And thanks, finally, to Hugh and Martin Kenrick who were kind enough to offer their time, hospitality, and family history to a couple of "nosy" Americans. We hope we have treated that history with the respect and care it deserves.

Abbreviations

AEU	Amalgamated Engineering Union
ASE	Amalgamated Society of Engineers
BDEA	Birmingham and District Engineering Trades Employers' Association
CEC	Children's Employment Commission
CIHMA	Cast Iron Hollow-Ware Makers' Association
DM	Directors' Minutes, Archibald Kenrick and Sons, Ltd.
EEF	Engineering Employers' Federation
LAB	Ministry of Labour
MEF	Midland Employers' Federation
MRC	Modern Records Center
MUN	Ministry of Munitions
NEF	National Employers' Federation
PRO	Public Record Office
RECO	Ministry of Reconstruction

Chapter 1

Introduction

Tlhe Kenrick iron foundry dates from 1791, the year Archibald Kenrick I, a plater by trade, leased a plot of land in the village of West Bromwich on the outskirts of Birmingham. The firm produced cast-iron "hollow-ware" for household consumption and commercial building purposes. Traditional products of the trade included pots and pans, coffee mills, and the like, as well as general hardware such as hinges, door knobs, cast nails, and so forth. By producing quality goods and distributing them in both domestic and international markets, the firm grew to more than 1,500 workers, controlled several subsidiary divisions, and came to play a major role in this segment of the British metal trades. The company still exists to this day; its original clock tower stands beside a busy motorway.[1]

The extraordinary longevity of the Kenrick family firm begs a deceptively simple question: how was all this accomplished? That is, how did the people involved—owners, workers, and others—go about producing and reproducing this organization? Choices were examined and decisions were made concerning the social and technical organization of work: how much work would get done? Who would do it and how would they be compensated? And what tools and technologies would be employed? Importantly, we want to ask: how were, what some would argue, the fundamentally divergent interests of capitalist and laboring classes accommodated at this firm? What other interests shaped these principle decisions regarding the social organization of work? What was the basis of "managerial authority" and how was it exercised? What decisions and conditions generated worker "consent" to that authority while others produced contestation?

These are fundamental questions about the social organization of capitalist work that we believe cannot be answered within the narrow parameters of economic or technological determinism. Clearly, such basic questions demand a conceptual framework that recognizes the daily life of the capitalist workplace is as much about politics, ideas, and identities as it is about exchanging work for wages, or the march of technological "progress." Thus, we begin with the assumption that the social organization of the workplace—its social structure and transformation through time—is shaped by historically specific political and ideological practices that influence and regulate struggles over interests and inequalities. These practices—thought of together as a "factory regime"—also shape the existence and character of possible collective action on the part of employers as well as laboring men and women.[2] We proceed by examining how the Kenricks used and, in the process, reproduced the unequal power and authority relations of class, gender, and age within the context of these factory regimes. These long-lasting, systematic, "durable inequalities," as Charles Tilly calls them, "arise because people who control access to value-producing resources solve pressing organizational problems by means of categorical distinctions."[3] Our challenge then, in studying the Kenrick case, is to examine how taken-for-granted assumptions about class, male superiority, and the social and cultural significance of familial relations were integral to the survival of the firm. Along the way, our analysis demonstrates how, as Heidi Hartmann asserted some years ago, "capitalism grew on top of patriarchy."[4] Yet, Hartmann might have added that in our attempts to understand capitalism, it has often been very difficult to *see* the patriarchy beneath it. To overcome this difficulty, we need a wide variety of detailed and historically grounded studies of capitalism that do not take patriarchy for granted—including studies such as this one—that focus on the individual firm.[5] Indeed, despite the considerable attention paid to the social, cultural, and economic changes that accompanied the arrival of capitalism in England, it is still the case that, as Patrick Joyce put it, "we lack an adequate historical account of that complex series of changes subsumed under the misleading heading of 'the industrial revolution.'"[6]

This book, then, is a study in social theory and historical sociology. Our historical analysis reveals three distinct factory regimes at the Kenrick foundry during the period from 1791 to 1922. We conceive of these regimes as successive forms of "capitalist patriarchy," which we take to be a social and economic system whereby adult men occupy positions of power over younger men, women, and children. We wish to underscore Judy Lown's position that women and children

in fact *cushioned* the effects of industrialisation for men. The ways in which women experienced industrialisation *enabled* men to secure a better deal out of the social and economic changes taking place. The processes by which this took place were by no means even, but they can only be understood by perceiving the essentially *patriarchal* interests at stake for men in both middle-class and working class positions. The shape those interests took depended greatly, of course, on the particular configuration of class and gender characteristics prevalent for men occupying different positions in the social structure (italics in the original).[7]

Therefore, we incorporate gender, as well as age, into our analysis of production politics at the Kenrick factory by identifying its effects on the labor process (in the hiring practices and the organization of work), as well as in the more generalized form of patriarchal power and authority as it shaped the character of particular factory regimes. To summarize, then, we see this book as making a contribution to our understanding of production politics, of patriarchy, and of the historical sociology of capitalism.

The Kenrick family firm represents a particularly important case because as it has an extended history that is relatively well documented and spans an important period in the development of British capitalism. Furthermore, the case permits an unusually close examination of human agency and action that is often lost in sweeping generalizations. The case study method is particularly useful in this regard, and, to date, few such studies exist.[8] Moreover, as we shall illustrate, the firm eventually played a central role in Britain's metal and engineering trades by negotiating and defining industry-wide arrangements concerning price structure, profits, and labor relations. Our intention, however, is not to argue that this case history is necessarily representative of other firms throughout this period; in fact, the characteristics we have just cited could support the argument that the firm was extraordinary. By analyzing the Kenrick experience, our intention is to provide a detailed historical account of the ways in which both capital and labor were confronted by the changing context of production and the strategies they undertook in defending and advancing their respective positions.

The Social Organization of Work in Historical Perspective

Assessing changes in the nature of work, specifically changes in its technical and social organization and cultural meanings, has been the focus of an enormous volume of scholarship in the human sciences. In sociology, post-World War II "industrial" sociology and the psychologically oriented "human relations" school gave way to the more theoretical and political "labor process" perspective. This work, much of it inspired by Harry Braverman's *Labor and Monopoly Capital*, focused on the coordinated set of activities and relations involved in the transformation of raw materials into useful products.[9] A key premise of the labor process view was the Marxist assumption that capitalism compels employers to continually transform the organization of production and workers' lives in pursuit of profit, and that workers often find compelling reasons to resist these efforts.[10] Focusing on the impact of the drive for profit on the craftsman, Braverman argued that, historically, worker creativity has been systematically undermined by the separation of thought and execution in work through "deskilling," the division of labor, the introduction of new technology, and the principles of scientific management. Others provided historical accounts of specific labor control strategies.[11] Still others focused on the social relations of production, emphasizing the generation of worker "consent,"[12] the creation of segmented labor markets,[13] and the impact of technology.[14] On the whole, labor process scholarship offered an important and lasting contribution to our understanding of the social organization of work.

Yet, with few exceptions, this literature painted an overly deterministic portrait of the historical transformation of the workplace—a workplace characterized by what Joyce has called the "inexorable unfolding of capitalist rationality acting upon a passive work force."[15] This view tended to overemphasize control and domination, and left little room for human agency, historical diversity, or questions about the *re*production of social relations. By highlighting the economic movement of production, much of this literature failed to distinguish the political, cultural, and ideational conditions that regulate these activities and relations.[16] Undermined by these failings, by the mid-1980s, the labor process literature lost much of its intellectual vigor.

Meanwhile, the discipline of labor history has experienced its own advances and challenges. Breaking away from both a political and a diplomatic history "from above" and the "old" labor history of male-dominated unions, the "new" social and labor history of the late 1960s offered "an expanded appreciation of the place of 'the political' in social life, which pulled analysis away from the institutional areas of parties and other public organizations towards the realms of 'society' and 'culture.'"[17] Indeed, a founding text in "cultural studies" was E. P. Thompson's *The Making of the English Working Class*.[18] In Thompson's account, British artisans developed a distinct, class-based political consciousness, not simply from their structural position vis-à-vis capitalists, but through their own set of cultural organizations, traditions, and ideas. Class was not a "structure" to Thompson, but a relationship, even a process. "Class happens," he wrote, ". . . when some men [*sic*], as a result of common experiences (inherited or shared), feel and articulate the identity of their interests as between themselves, and as against other men [*sic*] whose interests are different from (and usually opposed to) theirs."[19] Thompson's work inspired several generations of scholars to attempt to explain the processes and conditions of "class formation" and "collective action."[20] The "new" labor history "from below" offered a politicized discourse that was determined to uncover the authentic voices and experiences of everyday people. These social historians have, according to one recent assessment, "produced a rich and varied portrait of the European working class that has yielded new understandings about collective action; democratic and socialist political movements and ideological developments; working class culture, sociability, and leisure; household structures and their formation; changes in the content of work and the labor process; and the efforts of individual labor movements and leaders."[21]

Despite these considerable accomplishments, labor history has experienced its own revisionist "crisis."[22] Seemingly isolated, but not immune, from the influences of "post-structuralism" and the "linguistic turn" that has swept the human sciences, the new labor history now faces significant challenges. New interpretations and empirical studies that emphasize the role of discourse, language, and the social construction of meaning have offered compelling reasons to question some of the field's most fundamental tenets and assumptions. Some have called for a radical deconstruction of taken-for-granted categories such as "worker," "work," "wages," "skill," "men," "women," and even "class" itself. Indeed, some seem to suggest that the very concept of "class" may only be applicable in discussions of economic relations and does not apply to social and cultural relations beyond the point of production. These challenges—including some from the most respected names in the field—have questioned, among other things, the habit of attributing working class formation and resistance to the decline of subsistence living, the growth of wage labor, or the progressive loss of control of the labor process by skilled craftsmen.[23] This so-called "proletarianization" thesis has functioned, some say, as a theoretical and empirical "straitjacket" that unnecessarily limits our understanding. As William Sewell stated, "As a consequence, [labor historians] have paid insufficient attention to the profoundly uneven and contradictory changes in production relations, not to mention the role of discourse and politics in labor history."[24] Doubts have been raised about the "overly linear and deterministic" application of these developments,[25] the focus on incidences of conflict rather than acquiescence, and the casting of the skilled, male artisan as the heroic figure in the narrative of working class history.

Indeed, some of the most compelling criticisms of the new labor history have come from feminist theorists and writers of women's history. These scholars have attacked the field not only for leaving women workers out of labor history, but also for adopting and reproducing rather than replacing "a whole series of conceptual dualisms—capitalism/patriarchy, public/private, production/reproduction, men's work/women's work—which assume that class issues are integral to the first term of each pair and gender is important only to the second."[26] Since capitalism arose in a world where social life was, and continues to be, organized on the basis of presumed differences between men and women, gender has *always* mattered. The mistake, it would seem, was to conjure a classed but ungendered capitalism in the first place. From this revisionist point of view, "gender" is "a classificatory system that depicts the differing positions of women and men in society. It is a system of meanings articulated in practices that position women and men differently and that structures their lived experience in different ways."[27] It is therefore a social, cultural, and linguistic construction; a *process* of sorting people into categories and maintaining boundaries rather than reflecting something innate, natural, or "real." As some of the best feminist scholarship has maintained, "gender" is present even when women are not. Therefore, to "take gender seriously" does not simply mean including women in the picture, rather it entails accepting gender as a primary organizing characteristic of society. To fully understand the nature of social stratification, we must see that both capitalists and workers are gendered beings with, potentially, differing interests. Hence, we must analyze how gender hierarchies, authority relations, and possible allegiances are created and maintained, not only between men and women, but between powerful men and less powerful men, between adult men and younger boys, between boys and girls, and so forth. It also means that we must reject the aforementioned conceptual dualism that has permeated the new labor history. "If we do this," Lown declares, "patriarchal power can be characterised in terms of organizing and rationalising social relations based on male superiority and female inferiority which, *at one and the same time*, take an economic and familial form, and which pervade the major institutions and belief systems of the society" (italics in the original).[28] Clearly, we must understand the consequences of gender to fully comprehend the nature of class.

Class, Gender, and the Politics of Production

We think that it is critical to examine, in detail, the coordinated set of activities and relations involved in the transformation of raw materials into useful products, otherwise known as the "labor process." *How* employers choose their workers, *how* they organize their work tasks, and *how* they supervise and manage those tasks produce and reproduce class and gender distinctions. We also recognize that a structural class *relation* does not guarantee class *formation* or conflict. That is, we do not assume that "workers" naturally, ontologically, or ultimately have one stable, shared "interest" that motivates their actions; we argue that worker interests are unstable and fluid. We take up the challenge presented by Sonya Rose who asserts, "What needs to be made problematic is how interests are created. We cannot assume that they are inherent in social conditions. Interests are constructed through politics, and these politics create collective identities. . . . Unity or solidarity is a fragile accomplishment, and it must be explained by showing how it was accomplished."[29] We take the position that what workers want or need depends on who is speaking to whom, whom is spoken about, as well as where and when the speaking takes place.

Under the capitalist labor process, the transformation of potential labor power into useful labor involves a degree of coordination and control over worker effort by owners/managers. In order to generate surplus profit, the labor process must ensure this continuous transmutation. The accompanying social relations of production may be, in varying degrees, contentious or consensual. What we need to understand is the set of *mechanisms* by which these social relations are *re*produced and to uncover the contradictions that undermine and change them. We seek therefore to expand the sphere of production to include the political, cultural, rhetorical, and ideational conditions that regulate and shape struggles at the point of production. Here we draw on the work of sociologist Michael Burawoy.

Put simply, Burawoy's point is that a constant cannot explain a variable. Burawoy wondered how it was possible to explain empirical variation in working class resistance with reference to the social organization of work when, in many cases, we observe variation in resistance but no variation in work. For example, how can we, without completely abandoning materialist fundamentals, explain why in two cotton-spinning factories organized in more or less the same way, workers in one factory are unable, or unwilling, to resist the exercise of managerial control whereas in the other factory they are able, or willing, to do so? In his work, Burawoy criticized labor process theory for failing to distinguish, in any factory, between the particular organization of tasks required to get a job done (e.g., cotton-spinning) and the different ways managers might try to get workers to *do* the job (e.g., by coddling them versus beating them). He has argued that it is crucial to separate the labor process from what he calls "the political apparatus of production," or "factory regime," because these two components of production vary independently, and because variation in factory regimes—not the labor process—is what should explain variation in working class resistance. This cultural and political moment of production is analytically distinct and causally independent of the labor process. A factory regime, then, refers to the overall political form or system that mediates, regulates, and shapes struggles between or among workers and owners/managers, and that, in turn, may shape the *potential* collective interest and capacity of those workers or owners. The notion of a factory regime, we think, offers a way to balance our concerns with both the material and cultural dimensions of production. Following Burawoy's lead—yet offering a more explicit account of the role of gender in factory regimes—we argue that the character of any one political regime of production is determined by a changing set of historically specific conditions.[30] By treating each of these conditions as "independent variables," we can illuminate their individual effects on the form of the factory regime.

The first condition is the *structure of industry* or the degree of interfirm competition in the marketplace. Marx argued, for example, that during certain periods, the competitive pressures of the market may threaten the very existence of the firm. Under these circumstances, capitalists often respond with the introduction of new technologies or the intensification of work.[31] Braverman, on the other hand, contended that limited competition enables capital to confront the control of the labor process by the recalcitrant craft worker. In any case, it is clear that varying degrees of competition enable or constrain the choices and strategies undertaken by both capital and labor. The second condition relates, specifically, to the *organization of work* itself. Here we retain Marx's three-step process regarding the transition from the "formal" to the "real" subordination of labor to capital.[32] In the first "proto-industrial" phase of domestic production, workers generally own and control the instruments of production but find themselves vulnerable to exploitation by merchant capitalists. The second phase of formal subordination occurs as labor is brought under one roof. In this phase, capital owns the means of production but the worker still retains control over the labor process. The real subsumption of labor follows the separation of conception and execution of the work task by capital, and machines are used extensively to "deskill" workers who become an objective part of production.

The degree of *worker dependence* on the sale of their labor power constitutes the third condition of the factory regime. In the early proto-industrial phase, access to subsistence agriculture may have checked the reliance of rural craftsmen on merchants. Typically, what Burawoy calls a "despotic" regime is founded on the complete dependence of workers on wage employment for their livelihood. Under these conditions, the reproduction of labor power is dependent exclusively on a worker's performance in the workplace. In later phases, two forms of *state intervention* break that dependence and define the fourth condition of the political apparatus of production. Here, ". . . social insurance legislation guarantees the reproduction of labor power at a certain minimal level independent of participation in production" and effectively institutes a minimum wage. Moreover, ". . . the state directly establishes limits on those methods of managerial domination which exploit the dependence of workers on wages."[33] Compulsory trade union recognition and collective bargaining and grievance machinery offer workers some protection from arbitrary despotism. As Burawoy characterizes this change, ". . . workers must be persuaded to cooperate with management. Their interests must be coordinated with those of capital. The *despotic regimes* of early capitalism, in which coercion prevails over consent, must be replaced with *hegemonic regimes*, in which consent prevails, although never to the exclusion of coercion."[34]

Related to unionization and the structure of the marketplace, we see, fifth, the existence of *labor and capitalist organizations* as another crucial element in determining the character of any factory regime. Unions attempt to represent the collective interests of their members by demanding higher wages and improved working conditions, representing workers in disputes with management, organizing strikes, regulating member skills and qualifications, and, in general, providing a check on managerial discretion and power. Likewise, employer associations often limit competition among firms, fix prices and thereby ensure each others' survival, provide more stable markets, and negotiate regional or industry-wide agreements (or disagreements in the case of "lock-outs") with unions. Our sixth condition focuses on the local and regional geography of *labor mobility and opportunity*. We argue with Alan Warde that ". . . the power of the wage relation must be recognised as a factor in the maintenance of industrial discipline."[35] Prevailing local employment opportunities and relations, labor supply, as well as customs and traditions, must be considered as crucial in shaping struggles at the point of production.[36] Finally, we consider the *dominant ideological and discursive frames* that surround the relationship between capital and labor. We refer to the cultural and political discourse where "class" and class relations, social categories, and political identities are symbolically and rhetorically constructed. The work of the post-structuralists tells us that it is critical not to take language—the primary medium of culture—for granted. By examining the discourse of the principal actors, we ". . . will be in a position to determine the inclusions and exclusions that are central to the creation of political identities, and to the making of solidarities."[37] Evidence of this is found in the media, political speeches, statements and pamphlets by leading manufacturers and labor leaders, and in the diaries, letters, and papers of individuals.

While these seven conditions shape the character of the factory regime, and thereby the interests and capacities of the actors involved, such conditions should be recognized as outcomes of previous action. By conceptualizing these conditions in reflexive terms, we can better understand the production, reproduction, and transformation of factory regimes. Through the act of production, capital and labor produce and reproduce their own existence as well as the system of capitalism by applying the institutionalized rules and resources of recursive social practice. Yet within this process, actors may generate or confront situations that oppose these principles of reproduction; we consider these opposing forces contradictions.[38] It is within the dynamics of both social and economic contradictions that we locate the transformation of factory regimes. Basic contradictions emerge at the level of the state, the economy, and the firm.

For example, state intervention, undertaken in the interest of social stability, and, in effect, in the interest of *all* capitalists, may also come into conflict with the interests of *individual* capitalists. Economic contradictions include an oscillating business cycle that produces overproduction, underconsumption, and the centralization of capital. And unregulated competition among firms also generates backlash from workers as employers lengthen and intensify the workday in order to maintain profits and market share. In addition to contradictions that emerge from outside the immediate sphere of production, others may develop at the level of the firm which challenge capital's dominant position in the production process or, on the other hand, force labor to evaluate its role in that process, bringing about struggle and change. In that case, the *social relations* of production come into conflict with the *forces* of production (i.e., the power to transform nature through labor). In sum, changes in the determinant conditions of factory regimes are the result of shifting social and economic forces that alter the context of production. The character of a particular regime reflects these changing conditions as both capital and labor confront them in pursuit of their interests.

Capitalist Patriarchy Explored

Our historical analysis reveals different types of factory regimes and highlights the "variables" that determine the character of any one regime. We identify three distinct factory regimes at the Kenrick firm between 1791 and 1922. Our analysis examines their independent effects upon the regulation and mediation of daily production relations and on the existence and character of collective action on the part of both capital and labor. We conceive these regimes to be successive forms of "capitalist patriarchy."

The first regime, which we call "Patriarchal Despotism" (1791-1867), is covered in chapter 2. Under this regime, while Archibald Kenrick I, a skilled craftsman, was the reigning patriarch of his foundry, much of the work was done by adult male subcontractors who were paid according to an established price list for each piece produced. These subcontractors, in turn, hired young men and boys—mostly family members and kin—who were paid day wages and were brought into the workplace under the control and authority of these adult males. Therefore, the existing patriarchal social relations of the pre-factory period survived in this "early" factory. These circumstances encouraged petty-capitalist motivations on the part of these laboring subcontractors who worked under one roof but retained control over the labor process. Thus, our analysis of this regime highlights the formal rather than the real subordination of labor to capital. Direct capitalist involvement in the production process was minimal. Yet, while laboring subcontractors retained control over the labor process, they and their laborers had few options other than to sell their labor power, because they had no other means of subsistence; there was no state intervention in the form of labor legislation nor collective labor organizations to protect workers from arbitrary despotism; and competition led to the intensification of work and the extension of the workday. With rare exceptions, women and young girls did not work at the factory, but they performed unpaid domestic work at home in support of the household's wage-workers—usually men.

By contrast, under a growing "Paternalistic Despotism" (1868-1913), the owners were increasingly involved, both ideologically and materially, in the lives of their workers. We argue that this kind of paternalism was a legitimating ideology for a new patriarchal configuration.[39] Under this regime, described in chapter 3, the founder's descendants began to appeal to personal ties of dependency between themselves and their workers; this relationship was framed increasingly in familial terms. This paternalism was the political foundation of a new morality of economic rationalism, which, while emphasizing productivity, provided the ideological legitimacy for employers to challenge the remaining vestiges of skilled male workers' claims to control over their shops. Under this regime, these laboring subcontractors were seen increasingly as *dependent* "employees," rather than semi-*independent* producers; and their authority at work was overshadowed by both their employer's own patriarchal power and modest state intervention in the form of restrictive child labor laws. In addition, the employers provided basic "welfare" benefits to male workers and participated in and contributed to local charities, hospitals, schools, and other community institutions. This style of paternalism attempted to unite family, work, and community life, and to align the economic and gendered interests of capitalist society and the skilled male worker.

The potent effects of paternalism as a political apparatus were tested in the late 1880s when threats to profitability forced the Kenricks to seek new ways of organizing the production process. Laboring subcontractors, spurred on by an activist assembly of the Knights of Labor, responded to these incursions with the first strike in the history of the Kenrick firm. In response, the owners attempted to mechanize hollow-ware turning at the works and to dispense with their skilled, but expensive and increasingly troublesome, male turners. We argue that the conditions of this paternalistic regime shaped these struggles between the skilled men and the Kenricks over the employer's strategy to mechanize skilled trades in favor of capital, permitting the introduction of new technologies and "dilution" of the workforce with "unskilled" women workers. Paternalism and this tactic of mechanization worked together to reorder the gender hierarchy of the firm by emasculating skilled male workers and elevating the owner to the level of father/patriarch. Therefore, the real subordination of labor—the separation of conception and execution of the work task by capital—occurred with little resistance from labor. Our analysis also focuses on the importance of organized capital in this regime. By restricting competition among firms, the Kenricks kept prices relatively high, providing them with the capital for mechanization. Finally, although workers were still dependent on wage labor, they had less reason to resist the Kenrick's turn to machines because of new employment opportunities created by regional economic expansion.

Once again we found that the Kenricks solved, as Tilly suggests, "pressing organizational problems by means of categorical distinctions." The "pressing organizational problems" in question were the need to reduce the power of skilled male workers and to maintain profits. The "categorical distinction" they deployed to achieve these goals involved relying upon, and reaffirming, commonly accepted sexist assumptions about the differences between women and men. But what this strategy implied, of course, was treating women as a separate, and inferior, caste of workers.[40] This scheme was, ultimately, pregnant with contradictions. Since the Kenrick's paternalism did not readily apply to females, the more females the owners hired, the less effective paternalism was at binding the workforce to the workplace. Moreover, by failing to organize these women into a loyal workforce, the Kenricks left themselves vulnerable to having these marginalized and increasingly disaffected women workers organize against them. This vulnerability was exploited when, in April of 1913, the Workers' Union (WU) organized unskilled Kenrick workers in a dramatic two-week strike for union recognition and a scale of minimum wages. We state in chapter 4—contrary to the conventional interpretation—that the outcome of the strike at Kenricks, and of the Black Country Strike more generally, was not a victory for the "work people," but rather a victory for capitalist employers and working class men, and thus a victory for capitalist patriarchy. Indeed, what the story of the strike at Kenricks suggests is that patriarchy provided the point of consensus around which working class men and their employers could work out their (class) differences, resulting in both the preservation of capitalism and the reassertion of male authority.

Finally, a third regime, we call "Bureaucratic Hegemony" (1914-1922), is described in chapter 5. Under this regime, stepped-up state intervention—brought about in part by World War I—reduced labor's dependency on employers, which, in turn, reduced managerial despotism. Here we focus on the extensive involvement of collective labor and capitalist organizations in shaping the nature of factory politics. Under this hegemonic regime, consent prevailed over coercion on the shop floor.[41] Moreover, workplace struggles were increasingly embedded within a bureaucratic framework where wages, labor relations, the division of labor, and work performance were set out within the formal rules of rational-legal authority. This bureaucratic framework enabled and constrained the actions of *both* employers and employees. We examine how the unequal, sex-caste system already in place was further reproduced and governed by the state. We see this change as a move toward a form of "social patriarchy" and the beginnings of the modern welfare state in Britain. Specifically, we examine these conditions during World War I and the "Great Engineers Lock-out" of 1922.

Following is a comparative analysis of three production regimes within one firm in early British capitalism. This comparison reveals the role of these regimes in regulating the social relations of production and illuminates the contradictions that undermined and changed them.

Chapter 2

"Masters and Servants": The Foundations of Patriarchy, 1791-1867

In the latter years of the eighteenth century, a young man, Archibald Kenrick I (1760-1835), the son of an estate family from North Wales, himself apprenticed to the iron trade, entered into partnership with buckle maker Thomas Bolton. The two rented a shop in Hill Street, Birmingham. It was Archibald's second arrangement of this kind in a year; however, this one would last nearly four. While the buckle-making concern eventually fell victim to a change in fashion, despite a petition from local craftsmen calling for royal patronage, the partners had at one time hired six men and an apprentice, some paid by the piece, with the most skilled platers compensated by a weekly wage agreement.

Yet, as entries in Archibald's diary from 1787 tell us, the "habits of industry" were not common among the men. His frustration with their often "irregular" and sometimes drunken behavior is evident: "Paid the workmen who, having loitered their time away had little to receive, wanted to borrow, which I refused, excepting lending 1s. to one who had 5½d. due to him, and he returned it by not coming to work again for a week." This sort of experience lead Archibald to declare that, "Some plan must be laid down and strictly adhered to prevent the inconvenience of loss of time in the workmen: a present evil must be preferred to obviate a constant one."[1]

But it was not only the conduct of the men that troubled Archibald, for it would seem that the *geist* of entrepreneurship had not come easily to their master. Kenrick was often in turmoil over the apparent inconsistency between his own behavior and the tenets of his nonconformist religious upbringing. Early in 1787 he wrote: "I find myself as it were lost, not being master of what I am about, should exert myself more and act upon my own judgement or I shall always be subject to that of another." "March 11th: Trifled away the morning instead of reading some books to engage my thoughts to the duties of a rational being." "March 13th: Lay in bed til 8, determined as usual, productive of another evil—no resolution . . ." And finally, "May 18th: Continue to be shamefully idle in the morning, I am quite ashamed of my irresolution, it's a perfect disgrace."[2]

We see then, for masters as well as men, there was a struggle to adjust to new modes of social and economic life—a life in which the meaning of self, time, ownership, authority, and work and its products would be increasingly contested. This chapter begins to chart the course of these lives through the historical transition from "manufacture" to "machine capitalism," to examine the social relations implied by that transition, and to uncover the production regimes that regulated those relations. Kenrick himself would eventually embrace the "spirit of capitalism," in part, spurring his transformation from artisan-type small master to *petit bourgeois* small manufacturer.[3] But what of the workers? What would their equivalent transformation be? In time, they would change from relatively independent rural craftspeople to an industrialized proletariat, but this would take several generations. Despite his rather ominous desire for a "present evil," Archibald's vision of a timely, disciplined workforce was premature and his attempt at "rational capitalism" was frustrated. For the rhythm of production was, quite literally, in the hands of labor, and, for the time being, it would stay that way. But given the demands of emerging markets, competitive pressures, and lean profit margins, how could work be organized more efficiently? To answer that question, we must consider how metal working had traditionally been carried out in the region. Let us set aside the story of Mr. Kenrick and his workers for the moment to examine those customary practices.

Metal and the Midlands

The story of the city of Birmingham and its environs, known more commonly as the West Midlands, is, in one sense, an allegory of modernity. A crucible for the development of industrial capitalism, the region's history reflects a constellation of conditions and forces that would bring about a new social order. Yet, as seemingly brisk and sweeping as these changes would appear, we observe in this region a unique and convoluted passage, one characterized by great diversity in social and economic organization, custom, and culture. This variance provides fertile ground for pushing theoretical arguments to their limits and for offering a rich and fascinating context for historical study.

"Birmingham and District" is constituted by an approximately twenty-five-mile radius about the city in Southern Staffordshire and Northern Worcestshire counties. The district includes municipalities such as West Bromwich, Wolverhampton, Wednesbury, Smethwick, and Dudley. The western portion of the district, bounded by Wolverhampton to the west, Stourbridge to the south, West Bromwich to the east, and Walsall to the north, forms the area known as the "Black Country." Although the region as a whole is often referred to by this somewhat dubious title, the term originated in the coal field of this territory of South Staffordshire. A portion of this field has been characterized as a merger of thirteen or fourteen seams that form the "visible" and "thick" (ten yards) coal "for which the district was long famous."[4] So accessible was this coal that it could be exploited with little effort. "At Wednesbury, copyholders were mining from the fourteenth century in defiance of their lord. . . . Where coal could be dug so easily it was difficult for the lords of the manor to maintain a monopoly of mineral rights."[5]

Other natural resources were present in the region including ironstone, limestone, and fireclay. The ironstone, mined along with the coal, while not of the highest quality, did provide a ready supply for early iron-making; it would later be imported to the area in large quantities from the Forest of Dean, some sixty miles southwest of Birmingham. Other coal-related resources were the "thin" limestone, which was needed for flux in blast furnaces, and the "thick" or lower stone, which was converted into lime for building and agricultural purposes. Finally, fireclay from Stourbridge could be melted to produce glass, firebricks, and crucibles of various kinds.[6] In addition to these geological assets, the area forms a part of the main watershed of England that sustains a considerable number of streams and two main rivers, the Tame and the more historically important Severn and its tributaries. Since the district lies more than eighty-five miles inland from the country's main seaports, the Severn river evolved into a primary transportation and communication link throughout the seventeenth and eighteenth centuries. The Severn was more attractive as a commercial route since it was, from earliest times, a "public" river, free from the control of both private hands and the king. Navigation and utilization of local rivers was further enhanced by the construction of canals to link them with the water routes of other districts. The construction of the twenty-two-mile "Birmingham Canal" by Act of Parliament, in 1767, effectively linked the city with the ports of Bristol, Liverpool, and Hull. The extent of expansion of this system of canals in the region, from about 1770 to 1830, coincided with, as one author put it, ". . . a time of extraordinarily rapid growth in this town and in many other centres of industry, [and] might well be called the age of canals."[7]

But natural resources are, of course, only part of the story. It appears that the population of the region began steadily increasing in the 1560s with the "proto-industrial" parishes exceeding those agricultural ones.[8] W. H. B. Court estimates, in his authoritative history of the Midlands, that the total population for the three counties of Worcestershire, Warwickshire, and Staffordshire was as follows:

Table 2.1. Population of the Three Counties

	1670	1700	1801
Worcestershire	87,312	104,130 [141]	139,333 [189]
Warwickshire	83,389	98,725 [112]	208,190 [236]
Staffordshire	109,239	125,856 [111]	239,153 [210]

Source: Adapted from Court, *Midland Industries*, 20.
Note: Population density per square mile is noted within brackets.

These figures show that the total population for all three counties increased nearly 56 percent during the eighteenth century. This gain was marked, as well, by considerable increases in population densities (density per square mile in brackets).

The relationships between agrarian change and commercial capitalism, or countryside and city, are a crucial part of the early history of the Midlands. As early as the opening years of the sixteenth century, a large portion of the population derived its subsistence from the combination of farming and metal working, the balance of the two shifting somewhat from year to year. Evidence suggests, however, that cultivation was limited: "Communal openfield arable farming was not predominant in the area," according to Marie Rowlands, and, she goes on to say, "On the whole metalworking families who were possessed of goods in husbandry kept cattle and pigs, and ploughed little more than a few 'day works' of land in the open fields." Another historian asserts that there was "poor, inefficient agriculture throughout the district."[9] Data from the parish registers of West Bromwich for the first half of the seventeenth century indicate that more than half the occupations listed were nailers and buckle makers.[10] A sample of men's probate records for the parish covering 1660-1710 tells us that 41 percent were identified as "metalworkers."[11] The metal goods produced by these craftsmen included scythes, locks, swords, buttons, buckles, saddler's irons, and the ubiquitous nails, taking the least skill to produce.

Although larger workshops, employing hundreds of people and utilizing a division of labor and some machinery, existed in a few trades during the eighteenth century, the family was the primary unit of production in this "proto-industrial" phase of the Midlands (1550-1720).[12] Small, inexpensive, and easily constructed workshops were erected close to cottages. Rowlands estimates that no more than half a dozen people worked in these shops at any one time.[13] And, although generalizations are nearly impossible, it is probable that the "typical" household workshop was centered on an adult male craftsman, plying his trade with primitive hand tools passed down from his elders. This man taught the trade to his older sons or other male kin, and the craftsman and his apprentices were assisted by the less intense and "less-skilled," although crucial, contributions of women and children. Indeed, as Maxine Berg states, "Though women were the most important part of the proto-industrial workforce, the intensity labour from a woman with young children even if she was working at home was not likely to be high. It was the numbers of these women available for some work at less than subsistence wages, the numbers of their children who made some contribution to work, and especially the numbers of youths, who yielded both high labour intensity and high productivity, which made household manufacture so lucrative to merchants."[14] Toward the later half of the eighteenth century, women were employed in increasing numbers in the production of nails, buckles, toys, and so forth throughout the district. Berg also presents evidence of women in the Birmingham metal trades who either ran businesses themselves or announced their intention of carrying on the businesses of their deceased husbands.[15]

As a general rule, both the raw materials and finished products of domestic production were handled by the chapman, or wholesale ironmonger. The ironmonger, in turn, purchased his supply of iron from the ironmasters who owned the network of furnaces, forgers, and slitting mills (where bar iron was cut into rods) throughout the region. The typical ironmonger owned a warehouse and office to store raw materials and products and hired clerks to sort, weigh, and pack his inventory.

The character of the relationship between rural craftspeople and the ironmonger seems to have been quite varied. Some nailers appear more like "employees," debt to the chapman for raw materials being the primary consideration (one ironmonger, Henry Finch of Dudley, claimed in 1644 to have 100 men engaged for him,[16] while "The Crowely organization was employing 2,000 men in the Midlands by 1760").[17] Others, such as the more "independent" and highly skilled scythesmiths, would purchase rod iron directly from the ironmaster, circumventing the ironmonger entirely, and store and market their wares themselves. This suggests that the ironmongers dealt with the less-skilled segment of domestic producers, such as the nailers and locksmiths.[18] Importantly, however, no matter what the skill of the craftsperson, the conception, execution, and timing of the work in the shop was at their discretion regardless of their relationship with the ironmonger. In this sense, the ironmonger was "very much at the mercy of the honesty and industry or otherwise of the workmen."[19] Bargaining between the craftspeople and the ironmonger (or his rival) appears to have been quite common, because even the nailers, at times, had to be persuaded to work for certain rates. Given the simplicity of this production unit, the intensity of metalworking would fluctuate with demand, with agricultural seasons, and, later, within what constituted the "work week." That is, the "typical" work schedule evolved into one with little or no labor on Monday and sometimes Tuesday, followed by intensive bouts of work toward week's end, and culminating with a Friday or Saturday reckoning with the ironmonger.

We have then, in Marx's terms, the first phase of the *formal* subordination of labor to merchant capital and thus a capitalist mode of exploitation without a capitalist mode of production. During this early period of what Marx termed "manufacture," regionally clustered domestic producers crafted products for the wholesale ironmonger that were then sold in national and international markets. The mode of production of the Midland metal trades exhibited the following general characteristics:

1. the scale was typically quite small, employing little division of labor;
2. the immediate producers experienced limited competitive pressures, with reasonably high demand for their skill and products from numerous chapmen;
3. because minimal capital was required, it was easy for almost anyone to enter the trade;
4. there was little regulation of production by the state or collective capital;
5. it was anchored in the familial work process, so craftspeople had nearly complete control over:
    a. the labor process (conception and execution, hours and intensity of work),
    b. the disposition of finished merchandise (the most skilled "sold" *their* products to the ironmonger), and
    c. the means of production (land, house, shop, tools, etc.) were owned by the craftspeople; and finally,
6. farming provided an alternative, or supplement, to metal working as a source of subsistence.

By the close of the eighteenth century, however, the West Midlands had experienced the progressive disintegration of this "dual economy" and the emergence of a more urbanized "industrial" form. The demands of international markets, new construction, fashion changes, changes in raw materials and techniques, and a developing consumer market for new products generated increased demand for Midland metalware. Domestic workers responded by increasing their production. Moreover, while some lands were lost to the enclosures movement, greater amounts were simply consumed by the encroachment of buildings, coal pits, waste fills, and so forth.[20] Furthermore, land became valuable as a source of income, rather than as a source of subsistence, so families began to lease what land they might own for commercial/capitalist farming and to rent their cottages as residences or workshop space.[21] Burgeoning commercial centers, like Birmingham, drew immigrants from the surrounding rural villages seeking to associate themselves with the rapidly expanding industrial base.[22] These changes undermined the local, decentralized, kinship-based patronage of landowning and mercantile interests. As a result, power was vested increasingly in institutions controlled by an emerging urban class of business, bureaucratic, and professional elites.[23]

But these social changes also "proletarianized" the lives of these rural working people. In Rowlands's assessment, as early as 1760, "the workmen were less able to bargain with their employers than they had been in 1700. They were more dependent than before on a single source of income and therefore had to take work on any terms they could get it. The link between large populations engaged upon manufacture and the rising poor rate was already being pointed out by contempo-

rary commentators."[24] During this transition, the family began to lose its prominence as the basic unit of production as more workers left home each day for the
master's workshops. And, although many traditions and practices continued, their
meanings changed. The notion of apprenticeship, for example, began to lose its
more noble character, becoming ". . . merely a framework in which a new entrant
to the trade was hired on terms which enabled him to learn to be profitable to his
master."[25] Newspaper advertisements and surviving agreements for terms of work
show these changes. Apprentices were taken on as hired "hands," work was described as "constant," and tools were provided for the skilled and unskilled alike.
Rowlands notes the significance of these trends embedded in labor contracts: "The
emphasis is on full time work—the thirteen hours belong to the master. The workman cannot organise his work to profit from a variety of sources of income at
different times and seasons. He is not selling his products to the master but his
time and skill."[26]

But, while the once rural, independent craftsmen became increasingly dependent on manufacture for their survival, they rarely lost control over the labor process itself. And the budding capitalists, for the most part, had little incentive or
interest in contesting that control. Dependent on the craftsmen's skill to produce
high-quality products and lacking in any viable system of supervision, it was far
more advantageous for these capitalists to leave the coordination of production
and the direction and control of subordinates in the hands of the craftsmen. This
left few options—at least in the metal trades—for organizing work around anything other than some variation on this traditional arrangement.[27]

Patriarchy Put to Work

Given the history of domestic craft production in the region, it is not surprising
that entrepreneurs, like Mr. Kenrick, encountered great resistance when they did
try to tamper with traditional work habits. It appears that this lesson was not lost
on Archibald because he went on to *exploit* rather than *confront* the customary
control of the workshop by domestic craftsmen. With the dissolution of his buckle-
making partnership, Archibald embarked on a new venture, in March of 1791.
With the financial assistance of his father-in-law, Joseph Smith, he secured a ninety-
nine-year lease on a plot of land in the hamlet of West Bromwich about five-and-
a-half miles north of Birmingham. The property, located in the Spon Lane munici-
pal ward, was also advantageously adjacent to the Birmingham Canal and a road
that would later be a main thoroughfare. Kenrick was responsible for paying taxes
on the property and upkeep on any structures, and, within two years, he had erected
a foundry, workshops, a warehouse, and other outbuildings.[28] At his new foundry,
Archibald engaged in the production of articles of cast ironmongery, which were
quickly replacing those previously made of brass and copper. Otherwise known

by the trade name of "oddwork," these articles had domestic and commercial building uses and included bell-catches, hinges, doorknobs, handles, knockers, cranks, bolts, button-fastenings, flap-joint hinges, catches, and so forth. But in 1805, Archibald also commenced the production of cast iron "hollow-ware." The origins of this type of product date from a 1779 patent secured by a Birmingham craftsman, Jonathan Taylor, for ". . . an invention of casting oval-bellied cast-iron pots and [an]'nealing, tinning, and finishing the same and also . . . such kinds of round cast-iron pots and saucepans as are made with a head or rim round top."[29] William Hawkes Smith described the craft and skill involved in the process in his 1836 volume on manufacturing in Birmingham:

> the mould-case consists of four parts . . . the *drag*, the *two side pieces*, and the *top*. The *drag* is first filled lightly with sand, the vessel placed with its mouth downward, on the surface, the two *side pieces* (whose separation and junction is effected by means of truly cut brass slides and grooves) adjusted to their proper situation, and the whole held firmly together by pins and hooks. The inner edges of the side pieces precisely correspond with the section of the pattern and the casing, are filled up with sand, well rammed, by means of a tool resembling a pestle, the upper surfaces being levelled by a ruler, and powdered with dry dust. The upper division of the mould is then fitted on, and held in the usual manner by pins and hooks, and the filling proceeds. Embedded in the sand, in this portion of the mould, is placed an oblong piece of wood, called the *gett*, which, being afterwards withdrawn, forms the open mouth into which the metal is poured. The entire mould is then inverted, the drag taken off, and the interior of the pattern filled, the drag replaced, itself filled with sand, and the whole re-inverted. The case is then unlocked, the wooden *gett* extracted, the *upper part* removed, the coating of *dust* preventing the surfaces from adhering; the *side pieces* are gently drawn asunder by means of the brass slides, the two halves of the pattern severally taken off, and the *core* left standing in an inverted position on the filled *drag*. Its surface and the surfaces of the outer mould are smoothed where necessary with a piece of polished metal, and powdered over with charcoal dust, which improves the face of the casting. Lastly, an opening is made through the sand of the *upper mould* into the hollow formed by the *gett*, in order to admit the metal to the pattern. The mould is then replaced and hooked together, as completed.[30]

The process was completed as molten metal was poured into the mold, the resulting casting annealed, then "turned" with a lathe and file, and, finally, cleaned and polished.[31] Taylor's original patent for the procedure was later purchased by the firm of Messrs Izon and Whitehurst of West Bromwich (est. 1782 and, later, a long time competitor of Kenrick). Two primary types of these castings developed: the traditional "black" or unfinished, which was later succeeded by "tinned" or finished products. The process of refinement was later summarized, in 1865, by William Kenrick (1831-1919), Archibald I's grandson, this way:

The first improvement to be mentioned is one in the annealing, a process necessary to soften the cast iron before it can be turned bright in a lathe, preparatory to tinning. The method first practiced was very rude and ineffectual. The ware was packed in strong iron pots or pans, was piled up in the open air on a stage constructed of strong iron gratings placed side by side and end to end to any required extent. The whole was then covered over with coke, and the interstices, as well within the pans as without, were filled up with coal dust, to prevent as much as possible the access of air to the heated ware. The coke was then fired, and the pile kept at a red heat for about twenty-four hours. The absence of means for retaining and regulating the heat employed—in short, of conducting the operation safely and economically—is here evident. Mr. Archibald Kenrick . . . was the first to remedy these defects, by building an annealing oven. The annealing oven, as it was first constructed and afterwards improved at Mr. Kenrick's works, is an arched chamber, lined with bricks of Stourbridge fire-clay, 22 feet long, 11 feet 6 inches wide, and 7 feet 6 inches high; it has a fire-place in the middle, 5 feet wide, extending the whole length of the oven; it has flues in the walls and roof opening into the oven, and communicating with a stack high enough to cause a strong draught, which is moderated as required by means of dampers. Thick iron pans to hold the ware, 3 feet 8 inches high, and 2 feet 4 inches in diameter, are placed on each side of the fire-place, the flame and hot air from which envelope them completely in passing to the flues. Formerly, when the ware was sufficiently softened, the fire was allowed to burn out, and the oven gradually to cool. There were two drawbacks to the perfect success of this method, viz., the time occupied and the waste of heat consequent on lowering the temperature of the oven to a point at which a workman could enter it, and empty the pans. Both these defects were removed by the simple contrivance of placing the pans containing the ware on carriages, and running these in and out of the oven, on an iron tramway. This last improvement, it is only justice to state, was the invention of an ingenious bricklayer, named Moses Calloway. It dates from about 1817; the brick oven from 1807.

Mr. Kenrick improved the appearance of tinned hollow-ware by attention to the finish, and by substituting a stove-dried varnish for the black lead which had before been used as an outside coating. He also was the first to make saucepans with a rim or head round the top, an improvement claimed in Taylor's specification, but which had never been carried out—probably on account of additional difficulty in casting. Lastly, Mr. Kenrick's improvement in cast iron coffee mills, patented in 1815, gained for him the first name and the largest trade in this article which has since always been associated with the manufacture of hollow-ware.[32]

Clearly, Mr. Kenrick was an artisan craftsman. But he did not carry out his trade alone. In the early decades of the nineteenth century, evidence suggests that the firm employed somewhat less than one hundred adult men, young men and boys, and, likely, a few women who were involved in packing the finished products. A number of these workers were "outworkers," hired on a casual basis, their numbers rising and falling with the business cycle. But rather than attempt to retain the most skilled workers on a weekly wage contract as he had done in buckle making—and confront the problem of supervision and control—Kenrick contracted with skilled labor for products much like the arrangements made between the merchant and domestic workers in the past. Thus, the central feature of the firm's organization of production was an internal subcontract system. With this practice, Kenrick's subcontractors, or "servants" as they were referred to, were paid by their "master" according to an established price list for each piece produced. This arrangement was not unique to the Kenrick foundry; it was typical throughout the Midlands, and variations on the scheme appear as numerous as the firms in exist-ence. For example, some entrepreneurs merely rented space and equipment to independent "garret masters," with the "factory" little more than a shell, and the capitalist simply the owner of "the capital."[33]

The Kenrick foundry exercised somewhat greater control over its subcontrac-tors, because they did not rent space nor did the turners pay "lathe money," the relatively common custom of compensating "employers" for power and light. Importantly then, subcontractors at Kenricks owned no tools, equipment, or in-ventory. And yet, while these men were more numerous and less powerful than some, they had considerable control over the timing and execution of the labor process derived from their authority and skill in their shops. The legacy of this form of domestic production continued in the pattern of the workweek as the men "played away" on "St. Monday" and sometimes into Tuesday. Thus the workweek was often compressed into intensive three- or four-day periods with an average of 67.5 hours. The foundry was partitioned into uncovered shops that housed each of the subcontractors and reflected the four basic hollow-ware production stages: casting, dressing, tinning or enamelling, and finishing. An early record book en-titled "Stock: September 30th 1829 and June 30th 1830" identifies shop accounts, some by the first name of a subcontractor; "Robert's Shop," "Wright's Shop," "Robinson's Shop," Aaron's Shop," "Peter's Shop," and "Silk's Shop." Other sec-tions listed included The Foundry, Cupola Shed, Turning Mill, Grinding Mill, Blacking Shop, Large Counting House, Nail Warehouse, Sorting Shop, and so forth. Nineteen separate sections made up what was now Archibald Kenrick and Sons (the patriarch's second and third offspring, Archibald II and Timothy, had joined him in partnership). The practice of subcontracting is also evident in a "Hir-ing Book" dated 1835, which contains over ninety contracts between the company and workpeople covering a thirty-year period.[34] The majority include the follow-ing basic terms:

> William Woodhall engageth himself as a hired servant to Archibald Kenrick and Sons to work for them and no one else as a Turner for the new list of prices for the term of three years from this day of March eighteen hundred and thirty eight as witness his hand.

The page was then signed by both parties and a witness. There were a few varia-tions in the contracts, a substitution of occupation the most common, and included castor, bricklayer, enameller, carpenter, and so forth. Some of the agreements bound the servant for only one year, and even fewer paid by weekly wage, primarily enamellers. This was the case with William Lee who signed his agreement 22 July 1843:

> It is hereby mutually agreed between Archibald Kenrick and Sons and William Lee as follows—William Lee engageth himself to work for Archibald Kenrick and Sons and no one else as an enameller according to the custom of his employers enamel works for the space of one year from this time at the weekly wage of (18/-) eighteen shillings per week and to do his work faithfully and to the best of his ability.

And for James Ferguson as well:

> I James Ferguson hereby engage myself to Archibald Kenrick and Sons as a hired servant to work for them and no one else for the space of one year from Monday the 18th of March next at the weekly wages of twenty five shillings a week until a price by the piece is fixed upon. Archibald Kenrick and Sons agree to engage James Ferguson upon the above terms—West Bromwich, Feb 24, 1843.

It would appear the journeymen apprentices were similarly bound to the company, rather than to contractors, a practice that was unusual. A few contracts were for a "lad" whose agreement included his mother or father's consenting signature.[35] For example:

> Joseph Richards with the consent of his father Joseph Richards testified by his being a party hereto engageth himself as a hired servant to Archibald Kenrick and Sons to work for them and no one else for the term of three years from this twentieth day of March eighteen hundred and thirty-eight as witness his hand.

The hiring book contains numerous contracts of those having the same family name, as well as successive arrangements for the same person. Joseph Rock, for example, signed on as an enameller in July 1843 for the weekly wage of fifteen shillings, and he signed again, for another three years on the thirteenth day of March 1845, although he continued work "at his present rate of wages." Even though they were scribbled on the blank page of a book, these contracts were an important component of the prevailing social relations of production. Under the Masters and Servants Act, codified in 1823, it was a criminal offense for a worker to break his contract—either by leaving work without giving proper notice, neglect of work, or negligence—and the crime was punishable by a fine or a prison sentence of hard labor (a breach by employers was simply a civil offense). Research suggests that the Black Country was notorious for the widespread use of the Acts, and that they were ". . . used by the employers as both an instrument of industrial discipline and as a weapon to control the free movement of labour."[36]

For subcontractors working on piece rate, verbal agreements between them-selves and the masters would stipulate that the subcontractors would produce a certain number of items on the current price list by a specified time. The subcon-tractors, in turn, hired their own assistants and paid a daily wage to them to com-plete the agreed-upon contract. Given the existence of this "double wage" contract system, that is, the combination of both piece and day rates, a subcontractor's profit was dependent upon his ability to hire the right number of assistants to complete the job and to obtain the maximum productivity from these workers. Occasionally, such a margin was not realized within the terms of the contract, and the subcontractor had to borrow from the masters to pay his own employees. Led-ger entries entitled "Workmen's Debts" appear in the account books and were, at times, substantial. These were repaid in weekly installments to the masters.

The subcontractors at the firm were, without exception, adult males who hired their own workers. These were sometimes family members or relatives, and by all evidence, were exclusively young boys (at least for the first half of the nineteenth century). An individual shop might have as few as two assistants or as many as twenty, depending on the stage of production. Some women and young girls were hired directly by the masters, paid day wages, and engaged in lacquering, varnish-ing, drying, and later, japanning, as well as packing products.[37] The hiring of young people by subcontractors appears to have been quite common in the district. J. Edward White, in his report to the Children's Employment Commission of 1862, estimated that there were "20,000 to 25,000 children and young persons engaged in the manufactures of the whole district assigned to my inquiry," the youngest between five and eight years old. He goes on to state that a greater number of them "are hired not by the principals, but by the adult piece-workers under whom they work" and some, "especially the youngest, work in factories and workshops for parents or relations."[38] Testimony provided by local metal trades firms to the com-mission bears this out. Henry Loveridge, iron caster, stated that "I have two of my sons working with me. All have their own sons to work for them if they have any old enough." Twelve-year-old John Speke "used to work in a brickyard . . . with father," while caster assistant Job Carter, almost age eleven, worked for his brother. An owner of a Walsall foundry stated that the "great many boys [that] are em-ployed" are "principally Irish, or the children of poor parents. In many cases they work with their fathers." Among them was Benjamin Russell, nearly ten years of age, who stated that he worked with his father and two of his brothers.[39]

Fortunately, the report of the commissioners includes evidence collected by Mr. White on visits to the Kenrick firm, as well as testimony by Archibald Kenrick II and several of the lads.[40] These accounts offer a detailed picture of the hiring practices and the working conditions for young people at the foundry at midcentury. White, introducing his findings, stated that,

> The greater part of the many boys are employed in the foundries, large but low buildings, the remainder in filing and turning. . . . When I entered the large foundry

at 20 minutes to 2 it was fully at work, though only two-thirds of the dinner hour had passed. A young boy saying that a quarter of an hour was the time that he usually took for dinner, an elder boy, questioned by one of the employers if that was all, said that he thought that in the regular way they took about 20 or 25 minutes, not more.

718. *Mr. Kenrick*—Our hours are from 6 to 6 with a half-day on Saturday, and for most of the hands on Monday also, and the regular hours of work are scarcely ever exceeded. . . . If a man begins work as soon as he has finished eating, which perhaps he may do, as it is piece-work, the boys who help him must begin too.

We do not wish for boys under 12 years old, but to allow only half time to all under 13 would throw a certain amount of loss on the families of those who are now employed. . . . In our case all the boys, with one or two exceptions, are employed and paid by the workmen, who engage to complete so much work for a given price and find the labour.

Testimony was also given by a number of the young people:

719. *Charles Curley*, age 10—"Thread knuckles," i.e., put parts of hinges together, "pun" dust, take up scrap (waste metal), take out sides (of moulds), riddle sand, skim metal, i.e., take off the surface of the molten iron from the top of the pot just before it is poured, and clean the work. The other younger boys do much the same. Come at 6 a.m., or a little before, and leave at 6½ or 6¼ p.m. Meals in here. Breakfast at about 9, and begin work again as soon as it is done, usually about a quarter of an hour. About the same time for dinner. Get a wash every night at home. Another bigger boy works under the same man with me. Get 3*s.* 4*d.* a week. Was at school six years till coming here 10 months ago, and paid 4*d.* a week, and go on Sunday but never was at night school. Learned off the maps, reading, and sums. Can read . . . Can write . . . Four times 3 is 12.

720. *George Moore*, age 9—Am 10 next year, but do not know when. Work with father at the same work as the last boy. Was never at school except on Sunday. Do not know B, O, or A; A is Y.

721. *Thomas Ferrars*, age 20—"Run sides," i.e., put sand into moulds. Went to work at between 7 and 8 years old, blowing bellows and helping the man. Go to school on Sunday, but never was at any other school. Cannot read, but know the letters . . . Believe the Queen is a woman, but do not know what her name is, or if it is Victoria.

722. *James Ferrars*, age 17, brother of last witness—Went to school last Sunday. Did not before, because I had no clothes, and never was at any other school in my life, and have not been taught anything.

The report added, parenthetically, that "These two brothers are said to have had great disadvantages at home, and to be irregular in their habits. Both are squalid and feeble looking; and the elder, though over 20, looks quite a boy, and so thin that every rib could be counted, the shirt being half gone."

723. *Joseph Baker*, age 16—Work at nail-casting. The apron is to keep the mould from hurting my legs and dress. Have been at work seven years. Never at school except sometimes on Sunday. Cannot read. (Knows the letters).

724. *George Jones*, age 13—File and turn at a lathe. Have been at these works since 7 years old. Get 7*s.* a week wages, and 3*d.* for sweeping for a woman. Father cannot afford to pay for my going to night school. Can read a little (but imperfectly; e.g., "was" is "as.")

725. *William Stringer*, age 12—Blow bellows. Pulling a rope with my foot and hands. Cannot read. (Reads words of two or three letters, and writes from sound "boy" and "5.") Was at school till 10 years old.

726. *William Holden*, age 12—At school till 10 years old. Did multiplication.... (Reads tolerably.) Am about the best scholar in this shop. There are nine boys in it. One can read as good as me.

The subcontract system at Kenricks consisted of, at the bottom, a relatively large number of young boys, accompanied by a few men, women, and girls, all of whom were paid day wages as underhands, assistants, or warehouse workers. These people, along with apprentices, worked under the direct supervision of ten or twelve adult male subcontractors who, in turn, worked for the Kenricks under piece-rate arrangements.[41]

For the better part of the nineteenth century, hollow-ware was produced the same way it had always been produced. The basic technology consisted of molds and hand tools such as the hammer, anvil, file, and grindstone, and, later, lathes, which aided the process of casting and dressing the product. The Kenricks did introduce steam-powered machinery for turning lathes and driving other machines sometime between 1805 (the commencement of hollow-ware production) and 1812, replacing the horse-driven machines and hand-operated bellows.[42] But these innovations did not fundamentally alter the artisan nature of the work. As William Hawkes Smith stated in 1836, these advances were restricted to ". . . that alone, which requires *more force* than the *arm* and the *tools* of the workman could wield, still leaving his skill, and experience of head, hand, and eye in full exercise. . . ."[43] So, it appears that this addition had little direct impact on the operation of the shops or the labor process.[44] It is interesting to note, however, that during the early years the firm entered into several partnerships with inventors who claimed to have developed machines to replace hand tools or mechanize production techniques. The Kenricks advanced James Warren, a Middlesex gentleman, 500 pounds to complete work on a screw-casting machine that would be the basis for a new company, Warren & Kenrick. Five years later, Warren was still unable to complete the project, and, consequently, the Kenricks refused to maintain the partnership.

In 1840, the Kenricks also pursued a turning machine by James Hood. They initially paid Hood 60 pounds for "a certain invention lately discovered by him of a self-acting lathe for turning hollow-ware or generally anything to a pattern fixed on the lathe . . ." And they agreed that Hood would receive another 140 pounds under the condition that the machine "completed articles now manufactured by them may be turned in as good a manner as they are now done by hand" and that the firm hold the right to apply for letters of patent on the device.[45] But, alas, the quest was soon relinquished, because, as William Kenrick would later explain,

> The lathes used for turning hollow-ware are the ordinary round and oval lathes, the turning tool being held and directed by the workmen. Though more than one trial has been made to introduce a self-acting lathe for the purpose, the attempt has been abandoned on its appearing that no saving in cost could be made by the exchange of a simple lathe for the more complicated machine.[46]

Competition in the hollow-ware trade was relatively brisk during the first half of the nineteenth century, caused by the relative ease of entry into the trade, low capital investment, and, at times, "higher than 'normal' profits."[47] Although the specific number of firms engaged in the trade during the earliest days of the firm is not discernable, evidence indicates that a number of general foundries produced various forms of hollow-ware. By developing and adapting new techniques to improve the quality of their products, and by reinvesting their profits, Kenrick's foundry grew during this time. The number of employees fluctuated between 250 and 400 following vacillations in the economy. Overproduction in the economy produced a depression in 1830, followed by a deflationary period, which lasted, with the exception of 1836, into the 1840s.[48]

In response to this steady decline, eight of the largest firms agreed to fix prices in an effort to halt sliding profits.[49] In addition, the Kenricks began round-the-clock shift work, in the early 1840s, to keep the furnace alight continually in the newly built enamelling section of the works. According to the testimony provided to the Children's Employment Commission, in this enamelling shop, "three young boys from 11 or 12 upwards are employed for about nine or 10 hours in attending to the men."[50] By the 1860s, attrition left only sixteen firms in the area who identified themselves as makers of cast-iron hollow-ware.

Shaping Struggles: Patriarchal Despotism, 1791-1867

In the preceding narrative, we described the social context and production process at the Kenrick foundry from its inception to the 1860s. We can now summarize the implications of these conditions for the politics of production. A rural "dual economy" existed in the Midlands prior to the founding of Kenricks in the late eighteenth century. In this dual economy, merchant capitalists "formally" subordinated domestic producers. This "proto-industrial" regime "succeeded" because domestic craft workers were content to work for less than they might have because they could continue to raise animals and grow crops to feed themselves. This regime, however, led to population growth, to the accumulation of capital in the hands of merchants, and to commercial farming. And, as we saw, setting up shop in this context was relatively easy, providing a "training ground" for new entrepreneurs. But, ease of entry into the trade also generated competition and diminished returns, which prompted those very entrepreneurs to create or find new tools and new ways of organizing production. Enter Mr. Kenrick.

By the time he had commenced production in the Spon Lane district of West Bromwich, Archibald had already confronted the legacy of the recalcitrant, male artisan. Despite the fact that such workers had by this time largely lost access to subsistence agriculture, as well as ownership of the means of production, their skill and small numbers within the district afforded them considerable power. This left small masters, such as Kenrick, as Karl Marx put it, ". . . largely defeated by the habits and the resistance of the male workers" and unable to take *direct* advantage of the unskilled workers, particularly women and children.[51] So, just as merchants had earlier capitalized on the existing household mode of production, new entrepreneurs, such as Kenrick, exploited its patriarchal structure. As Hartmann stated, ". . . capitalists took advantage of this authority structure, finding women and children more vulnerable, both because of familial relations and because they were simply more desperate economically. . . ."[52]

Embedded within the internal subcontract system was a production regime founded on patriarchy that permitted the small master, like Kenrick, to utilize characteristics of handicraft and at the same time enjoy some of the benefits of larger, centralized production. By engaging subcontractors, Kenrick avoided the responsibility of organizing, supervising, and controlling the production process, and spread some of the financial risks to these subcontractors. As in "proto-industrial" production, the work of women and children was a "hidden" source of profit set within the larger wage system. As Berg put it, "Innovation did not displace the household workshop."[53] And, finally, since each subcontractor controlled only one part of the production process and struck independent piece-rate agreements with the masters, the subcontractors had few common interests and thus little incentive to associate. This production regime, then, took advantage of craft control—and the preexisting labor process—at a time when entrepreneurs depended on the skill of these men and their workers to produce quality goods for a competitive marketplace.

The social relations of production involved in the piece/day wage internal contract were reproduced by this production regime. Burawoy stated, "Anarchy in the market leads to despotism in production."[54] The conditions of this factory regime permitted labor intensification strategies in response to declining profits and "successfully" contained shop floor struggles and possible resistance. Adult male subcontractors, pressured by their production quotas and price lists contracts, had to intensify both their work and their assistants' productivity in order to turn a profit—especially during deflationary periods—since piece-rate prices and contracts did not necessarily fluctuate with changing economic conditions. As assistants were paid on day wages, the owners and the subcontractors had the incentive to "sweat" these workers with long hours and shift work. Hence, this regime reflected a patriarchal hierarchy that involved a form of codomination between the master and subcontractor and, in turn, between the contractor and his assistants.[55] Consequently, the system encouraged petty capitalist motivations on the part of these subcontractors or "little masters." The patriarchal authority of Archibald I and his sons, as well as that of working class men, provided the system of control and left the majority of unskilled young men with little choice but to engage in the process, learn the trades, and, if they were fortunate enough, move on to hiring their own assistants.

Chapter 3

With "Liberality and Kindness": The Genesis of Paternalism, 1868-1891

The decade of the 1860s set the stage for significant changes in the organization of work, and indeed, in the political economy of England. The often severe living and working conditions of the laboring classes were becoming increasingly obvious to all, and many pointed to the urban and industrial Black Country as a prime example.[1] In his report to a local industries committee in the early 1860s, W. C. Aitken characterized conditions in one of the Midland trades this way:

> The diseases common to the brass foundry trade are chiefly of a pulmonary nature, and arise, in the case of finishers, from the dust involved in filling the brass; and in casters, from the dust that arises during moulding, and the condensed fumes of the volatised [sic] zinc from the melted brass in the operation of pouring. Brass casters are almost unanimously said to be short-lived. . . . C. Turner Thackrah, who wrote in 1832, also clearly describes the form of disease . . . "it affects the respiration, and less directly also the digestive organs; it is attended with difficulty of breathing, cough, pain at the stomach, and sometimes, morning vomiting. In Leeds we did not find one brass founder forty years of age."[2]

Aitken went on to state that nervous and respiratory ailments affected 261 inpatients and 1,542 outpatients at the Birmingham General Hospital during the period from 30 June 1864 to 1 July 1865. Moreover, he referred to the statements of an operative of the "better class" who collected money for the expenses attendant on the deaths of children—150 women worked in the factory he was connected with—and he believed that ten out of every twelve children borne by the married women in that factory died within a few months of birth. The author also stated that in the united parishes of Birmingham and Aston for the ten years from 1851 to 1861, deaths for children numbered 19,292 for those less than one year of age and 15,221 for those between the ages of one and five. The population for the entire district in 1861 was just over 290,000.[3]

Generalizing from his own investigations of particular trades, Robert Baker, a factory inspector for West Bromwich, wrote in 1868,

> Some of the japanning factories are dirty and ill-ventilated, especially the rooms in which blank trays are stamped. The same remarks will apply to rooms where hollow-ware is turned. In fact some of these places seem scarcely fit for anyone to work in. In many foundries, too, there appears to be a system of ventilation which, while it lets in abundance of cold wind on the workers, fails in removing noxious vapors. A few works had the machinery, &c. well guarded, but in the majority a great deal of fencing was required.[4]

The surviving evidence concerning the conditions of work at the Kenrick foundry shows that there is little reason to assume that conditions there were dramatically better *or* worse than those in the district as a whole. Recalling Mr. White's inspection of the "large but low buildings" at Kenricks, he did comment that in one of the turning rooms, "I noticed one of the men with a respirator. I was told that these were supplied as a protection against the dust and fragments which fly off in considerable quantities, but that they were not much used."[5] William Kenrick offered his views in a paper published in 1866, stating:

> The condition of labour in the hollow-ware and associated trades is favorable to the workmen; employment, at least in well established businesses, is regular and well paid; the kind of work though often severe is not injurious, and workshops are mostly large and not overcrowded. . . . The wages of skilled labour average from 25s. to 40s. a week. It may be taken as an evidence of provident habits in this class of workmen under favorable circumstances that, of the number employed by one firm, one-fifth are freeholders.[6]

And yet, again, in the same volume in which Mr. Kenrick wrote, the editor, Samuel Timmins, stated this about the hollow-ware workers of the district:

> Their social condition would be up to the average of the mechanic class, but there is still a serious amount of intemperance among them, and they are very improvident. Their homes are sadly wanting in order and general domestic comforts. Much of this should be attributed to the ignorance in such matters of their wives, who are for the most part drawn from the factories.[7]

And just what were the social conditions of the average "mechanics class" of the Black Country at midcentury? The general conditions were such that by 1861 the population of West Bromwich had risen to 41,795, somewhat less than eight times the population of 1801. There were slightly more women in the town than men, while approximately 50 percent were under twenty years of age. The population tended to marry later; there was an average of about five people per family, while the death rate for infants and men in their prime was above the average for the nation.[8] By the 1870s, only 41 percent of the 9,500 children ages five to thirteen in West Bromwich attended school.[9] Dr. Ballard's Report of the Sanitary conditions of the District tells us that "the whole of the district is unsewered, the result is that slops and sewage running from the domestic premises . . . form stagnant pools in the channels and corners of the streets, and occasion in some cases abominable nuisances close to inhabited house. . . . Nothing can be more disgraceful than the condition of these places, which are utterly neglected, and where the inhabitants are compelled to live in the midst of extreme filth." Privies were universally inadequate, designed as ash-pits that were not cleaned often enough. And only 16 percent of the homes in West Bromwich received water from the Water Works Co. Others used three public wells and some private wells, "very many of which are very superficial and contain water obviously polluted with sewage and filth."[10] Still others simply went without.

According to George Barnsby's study of the district, to maintain a minimum standard of comfort in the 1860s, for a family with two children under working age, wage earners needed 28 shillings a week, with 14 shillings per week necessary merely to subsist.[11] It would appear that at least the skilled workers at Kenricks, virtually all adult men, were able to earn 28 shillings per week, and thus attain Barnsby's minimum level of comfort without help from other family members. However, Barnsby calculates the *real* wages of skilled engineering workers in the year 1866, for example, and finds that they had fallen 26 percent since 1850, while the cost of food rose 29 percent during that time.[12] He concluded from his extensive survey of life in the Black Country during the nineteenth century that things, "failed to improve before the 1890s" despite the so-called "mid-Victorian boom" of the 1870s.[13]

"Underletting"

Although attention was increasingly drawn to the general conditions of working class life, it was the treatment of children in the workplace that became a focal point of criticism by the public, health officials, labor supporters, and politicians. Included in his report to the Children's Employment Commission on manufacturing in the district, Mr. White wrote of his visits to the General and Queen's Hospitals of Birmingham. Here he not only uncovered evidence of wanton disregard for child safety in the workplace, but also evidence of blatant verbal and physical abuse that these children endured at the hands of the men who hired them.

728. *Joseph Hood*, age 17—Have been "taker-in" in a glass-house in Birmingham for four years, and worked the usual six-hour turns day and night, not working over often. The men are not very kind to you in a glass-house; they're rough brutes there. They beat you with iron things . . . and hit you about the head with them, and kick, and cuss, and swear at you; and they will do this for such little things, as, e.g., if you tumble down and break a glass. We do not tell the master, they would beat us worse then . . . Once . . . I saw a man hit a boy of about 12 on the back of his head with the blowing iron, which had some glass on the end of it, and cut his head open . . . We all catched it sometimes. They leathered us sometimes because we did something on purpose for fun, and sometimes when we couldn't help it.[14]

White added that "This boy is suffering from heart disease, brought on by his rheumatic fever. Looks very ill, and pants as he lies in bed. He is gentle and intelligent, a favourite with his doctor and nurse."

The next witness, fourteen-year-old Joseph Slater [730], suffered from heart disease as well. He complained about his job of "riddling sand" (part of the moulding process and a common responsibility of the boys), stating that he would work until,

my back used to ache so I could hardly bear myself, and when I went home I used to kneel down and put my head on a chair. Could not bear to sit up. Sometimes the men would kick we [sic], sometimes throw anything at us . . . One man kicked me on the bottom of the spine so that I could not bend my back for two or three days . . . The big boys used to beat us as well as the men, but they did not hurt us. My uncle was master (foreman?), and would not have let me be beaten if he had known it. Some boys was very often burned with the metal.[15]

Other witnesses went on to describe their traumas. One boy received a compound dislocation of his shoulder after he was caught in a drilling band and hurled "right up to the ceiling, and I fell out to the floor. There was no one in the shop to stop the machinery, only the little lad," his work-mate. The effects of lead poisoning were also described, and others showed how fingers and limbs had been smashed, sawed off, cut, and sheared.[16]

Blame for these conditions was laid principally on the system of "underletting," the term used to refer to the subcontracting system. Mr. White's report to the Commission put it this way:

I have been told by some employers that they "have no underletting; it is bad," as it is to the prejudice of the young, and diminishes the control of the principal. It is said also to tend to depress wages generally; the piece-workers, as well as the smaller employers, to make their own profit without exceeding the market price of production, employing younger and cheaper labour than the principals or larger employers would do, the personal interest and closer presence of the piece-worker enabling him to enforce an attention to work which the principals or larger employers could not secure. The practice is probably very convenient, but, speaking from my own observation, I have no doubt whatever that it is very much to the prejudice of the children and young persons employed here, chiefly by its removing the direct responsibility from principals to others who realise it less.[17]

Why White thought that the "principals" would necessarily take a more responsible position with regard to the children is unclear. Moreover, he seems to imply that the practice is somehow carried out without the consent of these owners. Yet, it is quite clear that the practice was "convenient" because it *did* enable the subcontractor to "enforce an attention to work which the principals or larger employers could not secure."

"Legislative Interference" and the Trades

Despite the mounting evidence that the continued employment of young people was detrimental to their health and well-being, many employers were, of course, ambivalent about making changes, and especially nervous about having changes foisted upon them. Archibald Kenrick II stated his position on the matter to the Commission of 1862. He said:

> I do not think that there is anything in the employment of the young in the manufactures of this district which calls for legislative interference; and this is the opinion of those manufacturers in the district with whom I have spoken on the subject. I do not see however, that, so far as we ourselves are concerned, regulations of the factory kind would affect us injuriously.

He went on to state, however, that:

> The practical effect of such a requirement in this district would be to shut out all under that age [under 13] from work altogether, for they are generally employed by the workmen themselves, and these I am sure would never have to do with double sets [i.e., having two sets of part-timers completing a full-time shift].

> If any measure at all were thought necessary, the best would be to limit the age at which children could enter upon employments, say 12, but to leave them free after that to work the full time. Education previously to that age should not be enforced, though it is very desirable that it should be had. I consider that to enforce it, e.g., as in Prussia, would be objectionable in principle as an undue interference with private liberties.[18]

Yet, by the late 1860s, changes *were* on the horizon. Production and output soared throughout the Midlands, culminating with the year 1872, which was described by one Birmingham paper as being "rarely equalled and never surpassed for its great and general prosperity."[19] William Kenrick estimated, in 1866, that the hollow-ware trade employed 2,430 workers including 1,370 men, 900 boys, and 160 women and girls. He went on to say that "About one-third of the whole number of boys may be taken as between the ages of nine and thirteen, and consequently liable to be subjected to the half-time provisions of the Factory Acts, should factory legislation be extended to this district."[20] Conditions were right for "enlightened" reforms, and, as William Kenrick anticipated, the metal trades, until this time essentially free from regulation, became subject to state supervision. The most significant laws having specific consequences for the trades were the Factory Acts Extension Act and the Workshops Act of 1867.

While the Act of 1833 had restricted the employment of children in specific industries, the amendments of 1867 came to regulate employment in *any* manufacturing process where fifty or more people were employed. These regulations outlawed the employment of children under eight years old, restricted those between the ages of eight and thirteen to thirty hours of labor a week, prohibited their nocturnal employment, and required them to have ten hours of schooling per week. Parents were liable to a fine of 20 shillings for noncompliance. The work of the young (between ages fourteen and eighteen) and women (eighteen and over) was also restricted to no more than twelve hours per day, within which at least one-and-a-half hours were to be allotted for taking meals and rest. Moreover, handicraft work was prohibited after 2 P.M. on Saturday, and all day Sunday.[21]

The relative prosperity of the day increased prices *and* wages as workers throughout the district began to demand—some through unionized representation—a share of the profits. The combined effects of the Extension Acts of 1867 and these wage demands had a dramatic effect at the Kenrick works. John Arthur Kenrick stated to an inquiry into the workings of the Factory Acts that, during this expansionary period, "The wages of half-timers have nearly doubled since the Factory Act came in, and of course all labour has gone up. We have had to increase our men's wages in some departments from 10 to 20 per cent in consequence of the great dearth of boy labour or the great prosperity of the country."[22] The shortage of child labor left subcontractors with few options concerning their workforce. They could employ continual shifts or "sets" of young workers, but this option promised considerable disruption in the shops. Alternatively, they could hire more expensive, older help, but this was no way to keep costs down. Or—as John Arthur testified to the Commissioners—they could, as his firm had done, substitute "idiot" for child labor on the night shift.[23]

The restriction on child labor was only one ingredient in a constellation of forces that would, in a few short years, sweep away work and leisure habits long associated with the Midland metal trades. As noted above, the Act of 1867 stipulated that no young person (between the ages of thirteen and eighteen) or woman could be employed at handicraft on Sunday, or after two o'clock on Saturday afternoon. This gave a boost to the nearly twenty-year-old Birmingham "Saturday Half-Holiday Movement," the creation of middle-class reformers, religious leaders, and certain segments of the working class who attacked the "evils" associated with "St. Monday." "Thus already the suggestion of a rationalization of hours had arisen," states Douglas Reid, and "It was a short step from this to the manipulation of the half-day by manufacturers so as to shift the leisure emphasis from Monday to Saturday; from the irregular unapproved 'playing away' to the recognized and much-lauded Saturday afternoon holidays."[24] A number of employers had complained in testimony to the Children's Employment Commission (1862) that the continued observance of St. Monday resulted in erratic work habits and unpredictable output. By contrast, in testimony to the previous Commission of 1843, St. Monday, a more or less accepted institution, was hardly mentioned at all.

In this context, it is understandable why labor's "Nine Hours Movement" (an actual ten-hour workday) of 1871-1872 encountered little resistance from employers, and the approximate fifty-four-hour workweek was widely adopted. While smaller shops generally resisted the reduction in hours, others welcomed the changes, particularly in larger firms where a reorganization of the workweek would impart some rationality to an otherwise "inefficient" domestic pattern of work. As Reid writes, "the Saturday half-holiday had been used as a sprat to catch a mackerel; a Saturday reduction in three hours (usually) in return for a Monday's labour of ten or eleven hours." As one employer put it, "Formerly the workpeople were apt to come in at all times, but the half-holiday enables me still more strongly to insist on regularity, and say, 'No, you have had your Saturday, and must be regular now.'"[25] Undoubtedly influenced by the Birmingham Saturday Half-Holiday Committee, some thirty of the largest firms in the area, including Kenricks, had granted the shortened Saturday "holiday" by the early 1860s. But, by then, workers at Kenricks had already agreed to come to work a half-day on Monday, laboring, on average, sixty hours per week. Recalling the testimony of Archibald Kenrick II to the Commission of 1862, labor was called by the bell in the clock tower at 6 A.M., left twelve hours later, and worked "a half-day on Saturday, and for most of the hands on Monday also."[26] By 1876 only a token of "St. Monday" was left. As John Arthur Kenrick stated, "Our hours are from 6 till 6, except on Mondays, when we begin at 7 and leave off at 5, and on Saturday we work from 6 till 1. We work 57½ hours a week instead of the ordinary number of 60."[27]

The Ties That Bind

The firm of Archibald Kenrick and Sons became a "private" limited company on 11 August 1883. By doing so the Kenricks took advantage of the protections offered by the incorporation clauses of the Companies Act of 1862, while keeping the firm's holdings under the complete control of the family. John Arthur, William, and George Hamilton Kenrick held 621 of a total of 625 shares issued upon incorporation, with single shares earmarked for their distant cousin, Frederick Ryland, who would become production manager, and three other relatives.[28] By dominating the board of Directors, the family consolidated and formalized its continued control over the business. This familial control provided the basis for an increasingly paternalistic relationship with the company's employees.

In the early days of the firm, Archibald I played the role of the charismatic artisan entrepreneur and "honorable master." Referred to by many as the "old governor," "tradition relates how he loved to get his apprentices round and give them sound advice."[29] Yet, given the notion of the firm's male servants as semi-"independent producers," the prevailing ideology in Archibald's day was one of laissez-faire and self-reliance, the masters having little responsibility for their servants' welfare and even less for the underlings that those servants hired. For example, the account "Workmen's Clubs" appears in the earliest ledger books, and we can only assume that these funds were insurance against illness and burial expenses. However, we can be equally certain that, in these first days at least, these small amounts were set aside by the masters by deducting them from the craftsmen's wages. Archibald I set up a works library and later, in 1840, his two sons constructed a school for their employees' children. Still, the required tuition and the fact that the school was not open at night made it less than practical, so much so that *not one of the employees'* children ever attended the school and the facility was later opened to any from the community who could afford to attend.[30]

With succeeding generations of Kenrick owners, however, we see the emergence of a "new paternalism," and thus the roots of a new political apparatus of production. Beginning in the 1850s, and coinciding with the Saturday Half-Holiday Movement—indeed an essential ideological underpinning of that campaign—was an attempt to create a community of interest between the masters and men. Joyce characterized these developments in these terms: "Labour was no longer regarded as a commodity. Aiming at work and hope, the worker was appealed to as a member of the whole community, no longer isolated in the insubordination and poverty that were once taken as the mark of immorality."[31] Employer and employee would join in a cooperative effort to create a social life and community institutions that would help all. This new ideology extolled the virtues of self-betterment for the working classes, but more, it was an artful blend of religious morality and middle-class values infused with philanthropy and good business sense. It was, indeed, to become the new morality of economic rationalism.

Nowhere was this new politics more evident than in the activities of the committee of the "West Bromwich Temperance and Educational Mission, for the Mental, Moral, and Social Improvement of the People" founded in the early 1850s with Archibald Kenrick II as President and Treasurer. The group would go on to form the Temperance and Educational Institute, with rental space and initial library provided by Kenrick. In a statement of the Committee written by C. T. Male, the Honorable Secretary, wrote:

> The exertions being made by the masters of the principal works in their district to improve the minds and the habits of their workmen providing books and periodi-

cals for them, and schools for their children, and evening classes for those at work, is in earnest of their sense of responsibility, and desire to do good, and that the necessary steps of progress to a better state of things are being taken all around us.

The ignorant and intemperate working men are a heavy tax on the sober and intelligent working men, who from their weekly wages, benefit societies, life assurances, and freehold houses are taxed by the vice and crime, disease, premature old age, death and pauperism which wicked conduct leads to. . . . Let us do our duty to our neighbors, and leave the rest to God.[32]

Second- and third-generation masters, such as Archibald II, Timothy, John Arthur, and William Kenrick, were not, like the "old governor," artisan entrepreneurs. Although they were technically astute, they were less the skilled craftsmen and more the bourgeois small manufacturers. As the organization grew, they spent less time on the shop floor and more time in the boardroom. And, well educated, they had much less in common with their workers than had the firm's founder. Thus, like other successful manufacturers of this era, these masters attempted to expand the domain of "consensus" beyond the point of production in order to forge a new, broader, social cohesion. This change was reflected in both industrial relations and the civic and philanthropic activities of these "new model employers."

The masters paid compensation for victims of accidents resulting from negligence on the part of the employers, and began to make regular contributions to the workmen's Mutual Benefit Society, The Birmingham Hospital Saturday Fund, The Workmen's Hospital Fund, as well as donations to the works' Cricket and Football Club. The lessons of "mutual respect" between Masters and Servants were cultivated at an early age. By the 1880s, apprentices at the works signed an "Indenture" statement that, among other things, attempted to tie their moral life and their work life to the firm. The document stated that the young man,

> shall faithfully serve, their secrets kept, their lawful commands and those of their Foreman and Manager for the time being everywhere gladly do. He shall do no damage to the said Company nor see it done by others . . . He shall not waste the goods of the said Company nor lend them unlawfully to any. He shall neither buy nor sell without the licence of the said Company. He shall not play at cards, dice tables or any other unlawful game. He shall not haunt taverns, alehouses nor absent himself from the service of the said company day or night unlawfully during the usual working hours; but in all things as a faithful Apprentice he shall behave himself to the said Company and theirs during the said term.[33]

The typical apprentice began work at age fifteen; and the agreement was binding until he turned twenty-one, during which time the young man was paid day wages that, with satisfactory performance, increased yearly. A parent cosigned the document in which the employer agreed to "At his own proper costs find and provide for the said apprentice during all the said term sufficient meat, drink, washing, medical attendance, clothing, and all other necessaries." Of course, benevolence had its price. The company had the lawful right to make deductions from wages for loss of time, neglect, or absence. "Gross misconduct," as defined within the terms of the indenture, were adequate grounds for dismissal. Often the last section of the Indenture certificate indicated the outcome of the agreement. For example, one reads "Alfred Wilkes has duly served his Apprenticeship during which time his conduct has been satisfactory, and he has proved himself a good workman." Another young man had his apprenticeship canceled.

Reaching beyond production relations, "new model" employers like the Kenricks were invariably active in the district's charity organizations and regularly attended local community and family affairs.[34] One local paper would later catalog the family's public service this way:

> Mr. J. Arthur Kenrick years ago occupied the position of chairman of the West Bromwich Board of Guardians; he was chairman of the West Bromwich Improvement Commissioners before incorporation of the town and for many years chairman of the Liberal Association. He is one of the oldest justices of the county of Stafford; has all his life been closely identified with various political and educational movements in the district; and is a Director of the Lloyd's Bank; Nettlefold's, and other important commercial undertakings. Mr. William Kenrick is a member of Parliament for the Northern Division of Birmingham, and has occupied the position of Councilor, Alderman, and Mayor, and in addition, has taken a deep interest in the Science and Art Department of Birmingham's higher education. Mr. George Kenrick has been for some years an active and hardworking member of the Birmingham School Board and has taken a great interest in the physical recreation movement . . . Mr. Fred Ryland has found time to do splendid work in the interest of technical instruction, the West Bromwich Institute, owning its inception, building and success mainly to his untiring efforts.[35]

Competition and Confrontation

A price slide in the heavy industries began in 1873, and the economic optimism of the past few years turned to despair. By 1876, the country was in the throes of the "Great Depression"; a deflationary period unparalleled in modern times.[36] Moreover, foreign competition from American, French, and German producers challenged Britain's traditional dominance in the metal trades. To contain competition among themselves, the British manufacturers formed the Cast Iron Hollow-Ware and General Iron Founders Association in April of 1872. With John Arthur Kenrick as chairman, the association aggressively sought to hold prices among its members. The thirteen or so associate companies divided themselves into three grades of makers, with Kenricks being one of two first-grade producers. These two firms controlled more than 40 percent of the total sales of the association. The group artificially set prices so that those of the lower grades could stay in business but not compete with the others.

Even though sales for the firm doubled during the period between 1870 and 1892, the deflationary conditions of the postdepression period had taken its toll; net profits as a percentage of capital employed, having peaked in 1876 at 26 percent, fell to less than 6 percent in 1886.[37] Faced with such numbers, increased foreign competition that threatened to disrupt a relatively stable domestic market, rising wages, and a shorter work week, the long-term profitability of the firm depended on reducing operating costs. State intervention and cries of "sweating" prohibited the labor intensification strategies of the past. Moreover, since the skill-based techniques involved in the production of hollow-ware had remained unchanged since the beginning of the century, such tasks could not be easily transferred to the semiskilled worker.

Confronting this dilemma, and despite the fall in profits, the Board of Directors decided upon incorporation to pursue an aggressive strategy of expansion of the company's physical plant (including the introduction of new machinery), as well as product diversification and acquisition. These tactics were financed through a combination of available profits, bank loans (John Arthur Kenrick had been a member of the Board of Directors of Lloyd's Bank since 1877), and personal savings. Between 1884 and the turn of the century, the firm nearly doubled its investment in the physical plant at Spon Lane, West Bromwich; and by adding new product lines and acquiring almost a dozen competitors and related companies, the company diversified its product lines.

The first machines adopted on a wide scale were stamping presses that produced seamless covers from a single piece of metal; the covers were patented by Fred Ryland in 1886 and soon were commonplace throughout the trade. More significant, however, was the introduction of hydraulically operated moulding equipment in the odd-work and hollow-ware foundry in early 1888, again introduced by Ryland. At a cost of £2848, these machines were designed to cut labor and raw material costs and to produce a standardized product. Accompanying the introduction of the machinery were redundancy notices for a number of skilled molders who were simultaneously replaced by less skilled men.[38] In response to these notices, the turners and tinners made the following demands:

14th Mar. 1888
Demands of A. Kenrick and Sons Employees
  I.   The reinstatement of the men recently discharged from the Foundry.
  II.   That they be set to work the machines in lieu of those since set on.
  III.   That an advance of 15 per cent on present prices be made to Turners and Tinners engaged upon machine made work, and compensation for extra labours rendered necessary by defects in such work.

With this statement the turners and tinners not only declared their support for their skilled colleagues in the foundry but also demanded a raise for themselves to compensate for the extra time needed to adapt their skills to machine-molded products.

A week later the employers flatly rejected these demands with the following statement:

> 19th March 1888
> *To the Employees of A. Kenrick and Sons Limited*
> Gentlemen,
> Having carefully considered your demands, we cannot disguise from ourselves that the real object which you have in view is to make the advantageous workings of the moulding machines recently introduced into our Hollow-ware Foundry impossible. Being firmly convinced of the advantages to be derived by use of these machines, both as regard to the economical making and better quality work produced, we feel ourselves constrained, having regard to the interest of the public and the trade, respectfully but resolutely to decline to accede to your proposals.[39]

All evidence indicates that no further action was taken by the men, the old price list prevailed, and the dismissed molders remained so.

Shortly after this episode, the turners at Kenricks joined a general movement of their colleagues throughout the trade to raise and standardize piece rates and improve working conditions. This movement was supported by officials of a local assembly of the Knights of Labor, the American masonic and labor organization, which was, according to one local paper, "gradually making themselves felt as an important body in this district." A considerable number of Kenrick workers were said to have belonged to the Order, so many so that the same paper estimated that the more than 900 employee members constituted a near "lodge" by themselves.[40] Early in 1888, the Knights, on behalf of the turners throughout the district, made representations to the hollow-ware makers. They demanded that:

> (1) a standardized price-list should be paid by all the manufactures of the United Kingdom; (2) that the masters should pay 1s per cwt. for swarf; (3) that there should be only one apprentice to every five journeymen employed, and that the journeyman should receive the benefit; (4) the day work wages should be paid at the rate of 8d per hour; (5) sundry minor prices for turning chucks.[41]

While it appears that a recognized price list for subcontracted hollow-ware products had existed in the Midlands for more than 50 years, many employers arbitrarily "discounted" prices, thereby, in practice, engaging in individual piece rate bargaining with their subcontractors. This, of course, created considerable variation in rates between firms and among the men, which they felt to be unfair. Furthermore, the subcontractors attempted to defend and advance their position with regard to several customary practices. They were requesting an increase in their portion of the sale of swarf, or waste metal shavings, as well as a reaffirmation of their control of apprentices (e.g., their pay and numbers in the shops). Interestingly, neither had been a custom at Kenricks. Moreover, the men wished to be paid a premium for adjusting the "chucks" that secured the hollow-ware as it was being turned. This was a task the turners had always done, but they were now demanding compensation for the additional time and skill allegedly involved in working the new machine-molded products.

The thirteen members of the now renamed Cast Iron Hollow-Ware Manufacturer's Association (CIHMA) responded collectively to these demands. They agreed, in principle, to discuss the idea of a more meaningful "standard" list (not a specific one) and the other issues, but only if the turners dropped their demands regarding the "privileges" of swarf and apprentices. It seemed, according to one newspaper account, that "the masters would not allow that the operatives had any right to products of manufacture, and were strongly of the opinion that the apprentices were better taught and very much better cared for under the control of the employers, who engage a competent foremen to teach the lads their trade."[42] Preliminary negotiations included the formation of a Conciliation Board and appointment of an outside arbitrator, Mr. Nigel C. A. Neville, Stipendiary of Wolverhampton, who solicited proposals from both sides. The men submitted the above demands, including a new list, while the employers, holding fast on "privileges," offered their own. The parties agreed, "That the award shall be accepted and acted upon without further question, so soon as it is published, and shall be binding on both employers and turners for twelve months from 29th January, 1889."[43]

On 2 April 1889, after considering the statements of both sides, the arbitrator proposed a new price list but made no mention of privileges. The men stated their dissatisfaction with the outcome in a letter to Mr. Neville, asking him "most respectfully" to justify "why he deemed a general reduction in wages necessary." His response was that he had given both sides "full consideration." While certainly not pleased with the outcome, the workmen "honorably accepted the award without solicitation," as the masters would later put it, and would abide by it for the next twelve months. Yet, shortly after working under the new price list, the contractors found themselves in debt. It was soon obvious to both the men and the masters that Neville had raised prices on certain articles, of which few were manufactured, and had lowered prices on articles in demand. A local reporter suggested that this mishap was perfectly understandable given the very complicated nature of pricing in the trade. Understanding the predicament, employers at some works, including Kenricks, forgave the debt, raised prices on certain articles, and even returned control of swarf and apprentices to the men, as was done at Clark and Co. This, however, contributed to even less uniformity throughout the trade, creating further ill feelings among the men.

In early January of 1890, the issues were again brought to the attention of the masters by the men since the contract period was soon to expire. The turners put forth their original demands. The masters refused to consider them, suggesting that the two sides engage Mr. Neville once more. This the men refused, citing their previous experience. Against the advice of the Knights of Labor, which called for negotiations, the turners issued a fortnight's notice to the Kenricks—and the dozen other mills in the district—that unless their demands were met a strike would occur. Neither side conceded; on 30 January, two days after the expiration of Neville's list, the turners struck.[44]

By the end of the first week of February, depleted stocks of turned hollow-ware threatened to idle 3,000 workers throughout the trade. At Kenricks, 63 striking turners were able to shut down the works and place themselves and 500 of their fellow workers "at play." The situation apparently created much friction among the various skilled craftsmen. The casters and other subcontractors "complained bitterly at the action of the turners . . . especially as it is so trivial a matter that the men are contending for." These men formed a committee to try to influence negotiations. The turners, however, remained steadfast in their position, despite rumors that support from the Knights was dwindling, which would leave the men without financial backing. In response to this hearsay, one turner who was interviewed stated that this was not so and that "the Knights were twice as strong as they were a year ago." "Then you expect to receive a considerable amount of support?" "Well, if a large body of men keep paying money into the funds, then there must be more available for a strike." As for the outcome of the strike, the craftsman said that "one firm in Liverpool had granted the men's terms and he believed that the employers of the Black Country would do the same." On the morning of the 14th, a number of Kenrick's tinners and casters, who were thrown out of work by the turners, "waited upon their masters works and asked them to advance a small sum to each until the matter is settled. The masters, however, refused to give them anything, considering it better to act with firmness in the matter, as the best means of bringing the strike to an end."[45]

By the first week in March, the CIHMA members met and agreed to "throw the doors open" on the 10th under the terms of the arbitrator's list plus 10 percent, but that the swarf and apprentices belonged to the masters. On the 6th, the turners met and voted to continue the strike; none "put in an appearance" on the 10th. On the 11th, the employers met again and resolved to keep the works open on the old terms, less privileges. Moreover, "At several of the largest of the works the masters are engaging men with the intention of teaching them turning and they (the masters) will take all responsibility of the spoilt work. Messrs. Kenrick, it is stated, are determined, rather than be beaten, to give up the hollow-ware trade altogether, but they believe that they will be able to do with out the strikers altogether in the course of a few weeks."[46] It was rumored that seventy men entered the Kenrick mill on the morning of 11 March. One of them was twenty-three-year-old George Brown who lived with his widowed mother-in-law, her five children, and two grandchildren at 208 Maria St., Spon Lane. He signed a contract with Fred Ryland to work "in the capacity as a hollow-ware turner or in such other capacity of any other branch of the business of the Employers as they may from time to time select for the space of six calendar months. . . . the Employer shall pay the workman the weekly wage of twenty shillings per week of 54 hours on the Friday each week."[47]

The striking turners had not anticipated this step. They assumed that the use of "blacklegs" would be self-defeating, as these unskilled men would spoil more work than they would finish. The employers determinedly pressed on with the trainees, however. Yet, on 13 March, John Arthur Kenrick, writing for the associated masters, made what appeared to be their final offer to the turners. They distributed a circular to every turner in the hollow-ware trade which stated that:

TO THE TURNERS EMPLOYED IN THE CAST IRON HOLLOWARE TRADE.

We very much regret that the opening of the Works engaged in the manufacturer of Cast Iron Holloware has not ended the disastrous strike of the Turners. We wish again to call your attention to the following facts:—

1st. The wages paid under the Arbitrator's award, produced more money for the Turners than did the old piecework prices ruling in 1888, (including all the Turners' so-called privileges). Wages for Quarter ending December, 1889, under Arbitrator's new award £4,630.30. The same period at 1888 prices (so-called privileges included) £4,6041.50. These figures show clearly that the Arbitrator did take into consideration the value of your so-called privileges. The same period at Arbitrator's award, plus 10 per cent; or the rate we now offer £5,093.35.

2nd. There are in our lists about 900 articles, of which we have given you 645 at the prices you asked, and the remaining 255 articles we and you agreed to arbitrate upon. Of these 255 articles, 172 were fixed by the Arbitrator at higher rates than ruled in 1888; 7 articles were fixed at the same rates: 53 articles were fixed at lower rates; 23 articles were hardly ever made, and consequently of these no record existed. Upon all these prices we have given you a voluntary 10 per cent.

3rd. You have in your correspondence with the papers, spoken frequently of the Nos. 5, 6, and 7 saucepans being reduced; you have never mentioned that the Nos. 8, 9, and 10 saucepans were increased. You are also perfectly aware that the whole of these sizes were those which your representatives said at the Conciliation Board were unfair, and which we agreed to alter, with the hope of averting the strike. To this offer you replied you would have nothing but your demands.

The terms upon which we have opened our works, and upon which we are prepared to re-engage our Turners, are:—

1st. That swarf and apprentices shall belong to the masters.
2nd. That day work shall be paid for at 8*d*. per hour.
3rd. That the proportion of apprentices shall not exceed 1 to 5, but apprentices engaged before January 29th, whether bound to the turners or to the employers, shall complete their agreements notwithstanding.

We would point out to you, that at two of the works the whole of the Turners have commenced work.

We shall now keep the works open, but shall make arrangements to engage hands to fill any vacant places, a decision which we much regret, but which we take in the interests of all our other employees.

> BALDWIN, SON & CO., LIMITED.
> T. & C. CLARK & CO.
> CANNON HOLLOWARE CO., LIMITED.
> THOMAS HOLCROFT & SONS.
> ARCHIBALD KENRICK & SONS, LIMITED.
> IZONS & CO.
> EDWD. PUGH & CO.
> JOSH. & JESSE SIDDONS, LIMITED.
> THOMAS SHELDON & CO.
> BEECH & RICHARDS.
> THE HILL TOP FOUNDRY CO.[48]

With pressure from the casters and tinners, diminished support from the Knights, and the apparent return to work by some of their colleagues, six days later the turners returned, en masse, to the shops and resumed production under the final terms offered, thus ending a five-week walkout—the first significant labor dispute in Kenrick history. For most of the men in the district, securing a 10 percent raise in the piece-rates of Neville's list was hardly a victory, particularly when weighed against the economic and symbolic loss of privileges. For turners at Kenricks, however, participation in the strike had successfully advanced their position—at least in the short term.

100 Years of "Liberality and Kindness"

Despite the contestation with the turners—or possibly because of it—the masters continued to advance their policies as "new model employers." On 14 November 1890, a serious accident took place in the Cover Tinning shop, killing George Sweeny, blinding another, and injuring five others. While a Coroner's Inquest found the incident to be, "Accidental death attributable to error in judgement on the part of Sweeny and exonerated the Proprietors from all blame," compensation was paid to the surviving widow. Following this episode, and at the urging of Fred Ryland, the directors approved the appointment of a trained nurse "to visit cases of sickness among the work people"; and later in 1891, they stationed two others in a joint position at the mill itself.[49] But it was the firm's centennial celebration in June 1891 that was, without exception, the most egregious display of the masters' paternalist ideology.

Early on the morning of 29 June 1891 four hired rail cars transported nearly 1,700 current employees, retirees, and their spouses to the Crystal Palace in Oxford.[50] As the trains arrived, the guests were greeted by a hand-shaking John Arthur along with seventy stewards who were handing out an eight-page program. The employees wore a distinctive badge and medal that they had received along with their engraved invitations. The medal bore the names of the Directorate on one side and a sketch of the works on the other. Across the badge were bars that indicated years of service to the firm: one silver indicated 2 to 10 years of service, 2 silver bars 10 to 20, and so forth. At least fifteen men wore the 3 gold bars, denoting 50 or more years of service, according to a reporter covering the celebration. After breakfast, the guests were treated to a performance by trained circus animals, tours of the grounds, and orchestral music.

After dinner in the concert hall, the employees presented the masters, sitting at the head table, with an ornately designed, illuminated address that read:

> The Centennial celebration of the establishment of Archibald Kenrick and Sons affords a fitting opportunity for the 1,200 employees to place on record our appreciation of the invariable fairness, justice, and courtesy which have contributed to the maintenance of the most cordial relations between employer and employed. We record with satisfaction the fact that we number amongst us many whose forefathers saw and assisted at the birth of this great enterprise and have helped to maintain in unbroken continuity those pleasant and mutually helpful relations. A combination of the Enterprise and Energy of the four generations of proprietors with the Skill and Industry of four generations of craftsmen, has crowned the undertaking with success, and placed it at the head of the Art with which the name of Kenrick must ever be associated. Long may that Enterprise and Energy, Skill and Industry, be combined to maintain the position thus attained. We are proud to be identified in some degree with this great industry whose productions are carried to the Four Corners of the Globe, and respectfully ask our employers to accept this expression of our gratification at participating in such an interesting celebration in the history of a private firm, and our recognition of the liberality and kindness with which such celebration has been marked.

Following the presentation, there were toasts to the queen, speeches filled with jovial reminiscences and personal accolades, while the "chord of paternalism" was, in R. A. Church's words, "struck again and again by employers, and work people."[51] Chairman John Arthur stated that the success of the firm was due to the fact that he, as well as his father, uncle, and colleagues had "tried to follow out the standard which his grandfather had laid down when he established the works. He tried to be just and to be fair—(loud applause). That had always been the standard."

William Kenrick, M.P., attributed the firm's successful history to their "reputation for honest and fair dealing—(applause), good value given to their customers, excellent workmanship, the result of those who had labored in the workshops with so much spirit and so much anxiety for the benefit of the concern—and good value given in every respect for their money." And he declared that a "friendly spirit which had existed, with hardly an interruption in so many years between employers and employed—(applause). No firm had been more faithfully served by its work people—(applause). He thought that they would agree with him that no firm in the trade had given better wages or better facilities and more healthy workshops; everything indeed, that made the condition and position of their work people better, more than this firm." Employee John White offered another toast to "Our Hosts" in appreciation for the chance to celebrate the centennial in this fashion and that, "A business which could stand the test of 100 years' existence must be built on a sound basis of integrity and justice. From fifty years experience he could testify that had been the policy of the firm towards their work people. . . . Their Chairman and the other members of the firm had always taken a personal interest in the welfare of their work people and to their honorable treatment might be attributed the accord that has always existed even when disputes between capital and labour had been general—(applause)."

According to the directors, this harmony was most challenged during times of technological change but that the accord had stood the test. In his speech, William stated that "the management of the business had always endeavored to keep abreast with the times by inventions, by improved processes . . . [and] it had been often found necessary to adopt, to a large extent, what were called labour-savings inventions, machinery, but they would see that these so called labour-savings operations had largely increased the ranks of the labourers . . ." He put it to them whether those labor-saving processes had not been working "in their interest as well as in those of the firms," and he hoped they would give the employers credit for,

> introducing any changes of that kind to take care that it should be introduced with as little hardship as possible. There must always be some hardship when they introduced new methods and new processes but they desired that this should be minimized, made as little as the conditions of the problem rendered possible. He might say in the presence of Mr. Ryland, who had so much to do in the last twenty-five years with introducing, inventing, and in superintending those processes, that they had been introduced with remarkably little friction and difficulty, and the fact that they had been accepted by the work people and zealously carried out was a great tribute to the fairness, good feeling and tact of Mr. Ryland, and was also a tribute to the sense and good feeling of the work people themselves—(applause).

And finally, Ryland himself stated that "they all hoped that if there was any spoil it would be equally divided between the employers and the work people—(applause)." He went on to assure them that in no time during his tenure at the firm,

> had any important improvements been suggested without the managers giving the most careful consideration as to what would be the effect to those to whom

the alterations would cause a change. Mr. Wm. Kenrick told them further that those changes sometime trod heavily upon a few. All he could say was that if that was true, they had to thank the employees most cordially for the way they had received the improvements. He had been most closely associated with them in the work of that old firm, and he did not think anyone could speak more strongly, or with greater thankfulness for the way in which they had always received any changes they thought fit to make for the welfare of all—(applause).

Yet few in the room were aware of the extraordinary changes to come. It would seem that, for the directors, the Centennial celebration afforded them an opportune moment to reassert the paternalistic underpinning of their "most cordial relations" with labor. As John Author stated in his Annual Report the month following the festivities, "The trip was most successful and gave great satisfaction to everybody and will, your Directors believe, tend to cement the friendly relations that exist between the work people and the firm."[52] By doing so, they were also laying the groundwork for another appeal to their worker force to accept, with "sense and goodwill," further "improvements."

It appears that the lesson of the turners strike of early 1890 was not lost on the directors. Undoubtedly feeling that they had been placed at a considerable disadvantage by the power displayed by the turners, and in keeping with their efforts to reduce costs, the owners had initiated plans to mechanize turning at the works. At their monthly board meeting of 16 April 1890—just one month after their "final offer" to the striking men on behalf of the employers—the directors, "Resolved that Mr. Ryland be instructed to make arrangements to make experiments for the purpose of bobbing work instead of turning it and that he may take premises elsewhere for the purpose."[53] Ryland successfully secured a patent for such bobbing lathes and turning machinery three months later.

At the December 1891 meeting, Ryland presented a report outlining the cost of the equipment and of the compensation to those who would become redundant by both dismissals and the substitution of the skilled craftsmen. After a series of tests, he concluded that the machines would save an estimated £4000 a year in labor costs. Ryland argued that, in addition to the machines, a total reorganization of the shop floor was necessary to reduce materials handling and yield optimal efficiencies. Yet, Mr. Ryland's indigenous, Tayloresque "scientific management" perspective was apparently lost on the majority of the directors. Only the machines were to be added, and, on the day before Christmas, 24 December 1891, notices were issued to twenty-four skilled turners.

This time, however, rather than simply dismissing the men as they had done in 1888, the employers had drawn up a compensation agreement that was first offered to the turners for their signature. The agreement stated that,

> Messrs Archibald Kenrick & Sons Ltd (hereinafter called the employers) having introduced into their works Hollow-ware turning machinery contemplate having

to dismiss from their employment a considerable number of journeymen turners now employed by them in the hand turning of Hollow-ware and they may have to put back to their former employment certain workmen who have bound themselves as Turners to the employers for a fixed term and who are hereinafter referred to as "bound men."

As the necessity of dismissing their workers as above mentioned arises from no fault of the part of the workmen themselves, the Employers are desirous of making such provision as is hereinafter mentioned for any workmen who may be so dismissed so as to assist them over the time that they will be out of employ whilst seeking new engagement. It is therefore agreed that . . .

1. Any Journeyman Turner who is being dismissed by the Employers owing to the introduction of Hollow-ware Turning Machinery shall be entitled so long as he is out of employment to the sum of twenty shillings per week until he shall have received the sum of Forty pounds upon the following conditions.
2. In the event of the Turner obtaining another situation in his own Trade the weekly payment of Twenty shillings shall cease and in lieu thereof he shall be entitled to receive a bonus of Twenty pounds down less than what he has received in weekly installments.
3. Should the Turner wish to permanently abandon his Trade giving the Employer a written statement that he is so doing he shall then be entitled to receive a bonus of Forty pounds down less what he has received in weekly installments or he shall be entitled in the alternative to receive Twenty shillings per week until the Forty pounds is exhausted.
4. Turners taking the benefit of this agreement shall at any time while in receipt of the weekly installments under this agreement return to their employment as Journeymen Turners with the Employers if so required and any Turners failing to return if so required shall receive no further benefit under this agreement.
5. Any bound man who shall be dismissed for the above cause will be reinstated in his old employment at the same rate of wages that he would be entitled to had he remained at his old work and upon his dismissal as a Turner he will receive a bonus of Ten pounds.
6. Any apprentice having completed his Indentures will be entitled to the sum of Twenty pounds on ceasing to be employed as a Turner.
7. Lastly this document is to operate as a separate agreement between the Employers and each person whose signature is appended to the Schedule hereto.[54]

The agreement appears to have covered all possible contingencies to the employer's benefit—even the rehiring of the men and the possibility that they might take the compensation and go to work for one of their competitors. On the other hand, the turners could receive as much as half an average weekly wage for forty weeks. The offer, considered generous by many, as well as rising prosperity throughout the region, appeared to have been adequate compensation for the men.

Presumably most found alternative employment while others chose to give up their trade altogether—either out of desperation or to take advantage of the "opportunity" to receive the £40. Those who accepted the offer signed an additional agreement that stated that they "would not at anytime hereafter during the period of seven years from the date hereof within a distance of 50 miles from the Town Hall of West Bromwich aforesaid be directly or indirectly engaged or employed or work or serve in the capacity of a Hollow-ware Turner."[55] Edwin Brookes, for example, swore off his trade and was certified by his new employer as a "collector," while William Lunn opened a small grocery store with his compensation.[56]

Apparently, the clauses of this agreement were taken quite seriously by the firm. In March 1893, after receiving fifteen pounds of weekly compensation, unemployed George Holloway agreed to resign his trade, and he received his remaining fifteen pounds. Yet *three-and-one-half years later*, on 6 October 1896, he wrote the following letter to the firm:

> Sir,
> Your letter to hand I have stopped work at Messrs Beach and Richards and I know the agreement and I know what it is to seek employment day after day for almost fourteen months and not get any I was compelled to go in the Hollow-ware trade again to keep a home for my family but by your request I am obliged to give it up at once.
> I remain, yours faithfully,
> Geo. Holloway[57]

Likewise, William Henry Bennett took his forty pounds thirteen months after the initial dismissal, but wrote this letter, which was received by the firm in the same month and year as Mr. Holloway's. It read:

> Dear Sir,
> I received your letter and see that you wish me to leave off working. I have seen my present employer about it and he tell [sic] me I cant leave without a weeks notice. Therefore I cant cease work before the 17th. I would have never had started there, only you know quite well my wife and children would have had to starve if I had not got something to do.
> I remain yours truly,
> W. H. Bennett[58]

Since no record exists of the company's initiating letter, we can only assume from the men's responses that the firm had, at a minimum, notified them that they were in violation of their signed agreements.

In the end, it would appear that, within the various clauses of the general compensation agreement, the sixty-four turners eventually made arrangements with the firm. Thus, all evidence suggests that the installation of hollow-ware turning machinery at the Kenrick mill took place with no organized resistance from labor.

Chairman John Arthur summed up the achievements of the year this way in his annual report in August of 1892:

> The Directors have much pleasure in reporting that the operations of the company during the past twelve months have been satisfactory and profitable and that the output at Spon Lane has been the largest ever recorded since the business was established in 1791. . . .
>
> But the most important and noticeable event of the year has been the successful introduction of machinery worked by girls for turning hollow-ware in place of the old method of hand turning done by men. Bonuses already amounting in the aggregate to £588.00 have been given to the displaced hands, and the introduction has been made without causing any friction or ill feeling and will in the future prove most helpful and profitable.[59]

Shaping Struggles: Paternalistic Despotism, 1868-1891

The last half of the nineteenth century was a watershed in the transformation of capitalism in Britain. Embedded in this period were the social forces that would strip away the remaining vestiges of domestic production and encourage the emergence of the "modern" factory. We witness then in the first one hundred years of the Kenrick iron foundry the historical transition from "manufacture" to "machine capitalism." And with this movement, we see conflict as well as compromise over the control and authority of the labor process.

In our second historical period of the firm, 1868-1891, the state intervened to restrict the employment of children and thereby shook the foundation of the subcontractors' patriarchal control and authority. These men could no longer drive their workers as in the past. "Sweating" came under attack by politicians, unions, and the public. As Craig Littler notes, "Few hands were raised to protect 'the slave-driving sub-contractor,' because new ideas, new methods, and new technology influenced employers to reach down for more control over the shop floor."[60] We see this at Kenricks where, by the late 1880s, threatened profitability forced the owners to seek new ways of organizing the labor process—a process long dominated by semi-independent subcontractors. At this time, Kenricks turned to mechanization, a strategy it had rejected in the 1840s, which was aimed at reducing both labor costs and the pivotal role played by craftsmen in the production process. In the later era, however, this approach had a number of attractive advantages: (1) it increased the productivity of labor; (2) the machines reduced the cost of labor by employing semiskilled rather than skilled labor; (3) the machines produced more perfect goods; and finally; (4) the machines removed the basis of the subcontractors' power in the labor process, as well as at the bargaining table. In addition, many employers in the trades attempted to take away from the subcontractors those traditional privileges that supported their power and authority in the shops. Labor's skill on the shop floor, which capital had harnessed in the past, had become, in this new environment, an obstacle to continued profitability. This contradiction reflected the tension between the employers' treatment of labor as a commodity, on the one hand, and their dependence on labor's ingenuity, on the other hand.[61] In reaction to these pressures, subcontractors like the turners at Kenricks, encouraged by local labor groups, engaged in a strategy of work stoppages with others throughout their industry to demand higher wages and to protest the encroachment of capital on their traditional domain.

How were the Kenricks able to later mechanize hollow-ware turning in the face of this potential collective interest? The answer, we argue, lies in the conditions of the paternalistic factory regime that would ultimately undermine and supercede the patriarchal authority of subcontractors, and overcome their resistance to mechanization. First, the limited intervention of the state removed a cheap source of labor from the subcontractors' shops. This subverted both the financial and patriarchal dominance of these subcontractors. The second important condition was the market structure for hollow-ware and Kenrick's leading position as a seller within that market. The oligopolistic hollow-ware industry created a relatively stable market that, in turn, provided the high and reliable profits from which the firm could fund welfare paternalism and finance the change from hand to machine turning, including payment of compensation to the displaced workers.

While profits at the firm roughly followed the oscillations of the national economy, the Kenricks were able to insulate themselves, through their manufacturer's association, from potentially destructive market fluctuations by regulating competition, prices, and output. From the firm's acquisitions, they derived income as well as depreciation credits and an expanded sales base. This resulted in more predictable profit margins and stable growth than if the firm had operated in a more competitive environment, and it contributed to the firm's appearance as a less risky investment when it attempted to raise capital from outside sources. Finally, in the absence of state-supported social security, subcontractors' and laborers' survival depended almost entirely on the sale of their labor power. This dependency left them more vulnerable to the hardline tactics of owners who were eager to assert their "managerial prerogatives" on the shop floor.

The Kenrick family firm emerged from the depths of the depression as a successful and prominent institution that was founded, in part, on a reputation for community service and philanthropy. The family's long-standing history as a major employer in the West Bromwich area where "generation followed generation through the works" laid the basis for the increasingly paternalistic relationship with its workers and their community.[62] Joyce summarizes these ideological and strategic developments throughout British society when he states that, "the English experience of industrialism deserves to take its place in the 'golden age' of paternal, dynastic, European capitalism between 1850 and 1875. It can indeed be argued that the English family firm of these years confronted the problem of size in a way that was more successful than variants of the military-bureaucratic model of management . . ."[63] It is this kind of paternalism which characterized the emerging production regime at Kenricks. This movement reflected a distinct shift in the attitude of employers concerning the welfare of their male workers. Ultimately, however, this new ideology hinged on a mutual understanding that, in exchange for its benevolence, the company could expect loyalty and hard work from these employees.

But the power of this new production regime went beyond a simple exchange of goodwill for diligence. This brand of paternalism was the political foundation of a new morality of economic rationalism that increased the power of employers at the expense of skilled working men. That is to say, paternalism provided the ideological legitimacy for employers to challenge the remaining vestiges of the skilled male workers' claims to control over the execution, hours, and intensity of work as well any lingering notion of "co-ownership" of the products they produced. The labor-centered observance of St. Monday fell victim to what Reid has called the "ideology of 'honest labour,' of the 'rational' use of time, of moral conformity to the steam engine's constant regularity."[64] This effort to take control over the labor process from workers was most clearly displayed in the rhetoric used by the employers in their attack on the "privilege" of taking swarf and controlling apprentices. Here the employers asserted in their negotiating demands that, "the swarf and apprentices shall *belong* to the masters" as they would "not allow that the operatives had *any right to products* of manufacture, and were strongly of the opinion that the apprentices were better taught . . . *under the control of the employers*, who engage a competent foreman to teach the lads their trade" (italics added).[65] In this discourse, the men are now "operatives" and "hands" who had no legitimate claim of "co-ownership"—not even to the residuals of production— and their skill and knowledge, as well, were called into question with regard to apprenticing new tradesmen.

Importantly, this shift of economic power from skilled working men to employer also transformed gender relations in and out of the factory. Those men who were replaced by women—and those who were left behind to work in an increasingly "deskilled" environment—were, to some extent, emasculated and superceded in a reformulated gender hierarchy that elevated the owner to the position of father/patriarch. It was within this context, which Church called a "stern yet considerate paternalism founded on a mutual respect," that the Kenricks were able to literally "buy off" the turners with a liberal compensation package—made especially attractive, no doubt, by a brief upswing in the economy, with its promise of brighter employment opportunities for the displaced men.[66] Moreover, the installation of turning equipment, as with the molding machines two years before, concerned only management and workers at Kenricks, unlike the work stoppage of January 1890, which affected the entire industry. Without the solidarity of turners throughout the region, Kenrick's workers were undoubtedly less optimistic about a victorious resistance to the new technology.

Actually, the resistance by male subcontractors at Kenricks appeared less focused on the installation of specific technologies and more on the particular issues of job security and payment structure. For example, turners and tinners had not challenged the introduction of molding machinery in early 1888, but rather the redundancy notices that were handed to their colleagues. They demanded that those men be rehired *to work the machines,* and claimed, moreover, an increase for themselves for crafting this new "machine made" work. Similarly, in 1890, while the turners' collective demands focused on the conditions of work, pay, and privileges, Kenrick's turners only sought gains in piece rates, ignoring the "preservation" of "privileges" that they, in any case, had never enjoyed in the first place. These facts illustrate the generally less militant tendencies on the part of skilled workers at Kenricks and provide evidence of the power of paternalism to shape working class resistance and capacities.

If paternalism was such a potent force, one might ask, how was the strike of 1890 even possible? Two points may help explain this apparent anomaly. The turners may have acted more out of altruism than militancy in supporting the demands of other workers in the strike of 1890, because the claims affecting the privileges of selling swarf and control over apprentices did not concern Kenrick's workers. Since it was the Knights of Labor who had initiated the demands leading to the strike, and since Kenrick's workers were highly visible in that organization, they might have felt compelled to participate. And, as Joyce observes, even when strikes occurred during the era of paternalism, they were more passive than combative.[67] The work stoppage of 1890 seemed to have this character; while there was definitely a dispute, hostilities were minor. As a local newspaper account paraphrased the workers' spokesmen during the strike of 1890: "inasmuch as although the men are on strike a most friendly feeling existed between the men and their employers, and there was no desire that any unnecessary friction should be created."[68]

In the end, however, the forces and conditions in existence under this emerging paternalistic regime proved instrumental in the demise of the system of subcontracting and its associated handicraft production. In retrospect, the resistance offered by the turners was mistakenly centered solely on payment structure. It would appear that they did not see a threat from the machines (i.e., they were willing to work the new machines but only within the subcontracting system over which they had considerable control). It was, however, the machines that broke subcontracting and the turners' position within it. Once installed, the turning machines, as well as other mechanized production equipment, had a dramatic effect on Kenrick's workforce. In 1876, approximately one-third of the workers at Kenricks were young boys, women, and children. Yet, by 1894, they comprised nearly 60 percent.[69] And by 1903, there was a complete disappearance of hollow-ware-turning apprentices at the firm. The social organization of work by the more traditional form of skilled male subcontracting was giving way to direct control of female, piece-rate machine "operatives" by an increasing number of company foremen. As Church noted in his study of the firm, "the process whereby the skilled handicraftsman retreated before the semi-skilled machine minder was well advanced."[70]

As we have shown, then, the Kenricks shifted from a patriarchal regime to a paternalistic regime, resulting in the real subordination of labor to capital, because broad political and economic changes altered the context of production, rendering the existing technical and social organization of work an impediment to the firm's survival. Thus, as Burawoy argued, the character of the factory regime began to change independent of the labor process. The real subordination of male, laboring subcontractors to capital at Kenricks, embedded within a strategy of mechanization, occurred *after* the state restricted work hours and child labor, *after* the Kenricks accumulated profits from their secure position in an oligopolistic (and colonial) market, *after* they had used much of those profits to encourage loyalty to the family in the factory and across the community, and *before* state-supported social security was available to provide workers with any alternative but wage-labor. These circumstances shaped the struggles around the mechanization of hollow-ware turning in favor of capital, permitting the introduction of the new technology. Against both Joyce and Burawoy, who argue that the bedrock of paternalism in the British textile industry was the real subordination of labor to capital, our study points to the possibility of just the opposite.[71] In the Kenrick case, paternalism provided the material resources and ideological cover needed to bring about the real subordination of labor to capital. In the next chapter, we see how the "success" of this strategy falters in the face of its own contradictions.

Chapter 4

"You Are Not Paid to Think": The Collapse of Paternalism, 1892-1913

At the 17 October 1894 Directors' meeting, Fredrick Ryland presented "a report on the substitution of machine turning for hand turning in hollow-ware and replacement of men by women" that he had undertaken three years earlier.[1] Ryland told the directors that the initial cost of buying-out (and buying-off) the craftsman, purchasing the new machines, and putting up a new workshop had been the considerable sum of £12,708, but he also added that the directors could now look forward to "savings in wages estimated at 47½%."[2] Ryland's invention and deployment of these machines shows how powerfully social factors shaped technological development at Kenricks. The "push" factors in the development of these machines were competition from other hollow-ware makers and the need to circumvent the power of militant craftsmen. The "pull" factor was the opportunity to make a better product more cheaply. Importantly, for our concerns, Ryland invented the machines with the *intent* to employ unskilled workers, particularly "female labor." Had Ryland worked out his estimates based on the use of unskilled men, rather than unskilled women, the savings involved would have been much less impressive.

Paying "female labor" less than the going rate for "workers" was an established practice in England by this time, and wages for unskilled female workers were set at about 50 percent or 60 percent of the unskilled male workers of the same age.[3] And there did not seem to have been much reason for it other than "custom." Reporting in their 1907 study of women's work and wages, Edward Cadbury and colleagues wrote that, when asked to explain the lower wages paid to women workers, "employers can usually give no other reasons for the actual wage than the fact that such and such a figure is what women usually get in Birmingham."[4] Clearly, Mr. Ryland was counting on this custom as he went about calculating the savings the Kenricks would achieve by hiring unskilled female workers.[5] Thus, from the mid-1890s on, unskilled male and female workers became a large and permanent part of the Kenrick work force, and thus played a major role in the Kenrick story from that point forward.

The impact of mechanization on the Kenrick workforce was evident as early as 1894, as shown in the table below. Whereas, in 1876, two-thirds of the Kenrick workforce were adult males, twenty-one years or older, this number had dropped to 40 percent by 1894, with a corresponding increase in females and youth.

Table 4.1. Kenrick Workers, 1894

Gender	Age	Number	Percent
Men	21+	489	40.0
Boys	18-20	77	6.3
Boys	13-17	302	24.7
Subtotal		**868**	**71.0 (Male)**
Women	21+	104	8.5
Girls	17-20	131	10.8
Girls	13-16	119	9.7
Subtotal		**354**	**29.0 (Female)**
Total		**1,222**	**100.0**

Source: Church, *Kenricks in Hardware*, 277 and DM, 19 December 1894.

Of particular note is the increasing proportion of female workers. In 1866, William Kenrick estimated that women and girls represented only 160, or less than 7 percent, of the 2,430 workers employed in the fourteen firms then making cast-iron hollow-ware in West Bromwich;[6] the table above shows that the percentage of females had increased to 29 percent by 1894.

Encouraged by the initial move toward machines and unskilled labor, from 1892 to 1913 almost every monthly meeting of the directors included a report about expenditures for new or refitted buildings and machinery.[7] And while some of this equipment was purchased to support the new product lines into which Kenricks was moving at this time,[8] the type of machinery purchased also indicates a continuation of the firm's deskilling strategy such that by July of 1914, on the eve of World War I, 33 percent (306 of 934) of the workforce was female.[9] The explicit purpose of deskilling the work of turners and others at the Kenrick foundries, as evidenced in the events of 1891, was to reduce labor costs and to place the owners in a more powerful position relative to their workers.[10]

The dominance that the Kenricks hoped to achieve could, of course, be maximized and solidified by replacing *male* workers with *female* workers; in addition to being cheaper, they would have expected women to be even less likely to cause trouble, particularly given the notorious difficulties organizers encountered getting unskilled women workers unionized.[11] Thus, when the time came to "degrade" labor, the Kenricks specifically sought out women workers for that particular dishonor. In contrast, as we saw in 1891, the Kenricks even fired their skilled men in a respectful, albeit paternalistic, fashion. Given what the Kenricks set out to accomplish, the shift toward female labor made sense; given sexist assumptions about the differences between men and women, the shift was possible. What the Kenricks hoped to do was harness preexisting beliefs about the inferiority of women, and put these beliefs to work in their struggle with male workers over control of production in the factory.[12] This strategy implied, of course, treating female workers as a separate, and inferior, caste. But this was hardly a new tactic in the history of English capitalism.

With the exception of textiles and clothing, where women were in the majority, almost all workers in England's nineteenth-century-manufacturing industries were male.[13] Yet, at various times and for a variety of reasons, English manufacturers did specifically attempt to hire female workers.[14] And this practice seems to have become more common across a wider range of industries as we approach the twentieth century. Based on her review of the literature and her own work on the nineteenth century, Sonya Rose concludes that, "Women comprised an easily identifiable and distinctive group of workers. They had already been identified as cheap labour prior to industrial transformation. When it was possible to do so, employers attempted to hire women in place of men, or they created new technology and designed it to be worked by women."[15]

What was true for English industrialists, in general, also seems to have held for engineering, and, specifically, the Midlands metal trades. Thus, Standish Meacham estimates that more than twice as many women as men entered engineering between 1891 and 1911.[16] A story entitled, "Midland Work-Slaves," in *The Sunday Chronicle* for 11 November 1911 began, "It is no exaggeration to say that Birmingham is the chosen home of the sweater of women's labor. This evil is at the root of all other social evils in the city, because to a very large extent cheaply paid women have displaced the skilled men artisans." Whether or not low wages for women were the root of all evil, the quote suggests that the process of replacing skilled men with unskilled women was generally recognized. So too, it would appear, in Manchester. In her study of women in engineering Drake quotes a trade union official testifying before a Poor Law Commission in 1908 saying, "the women are ousting the men in most trades, including the iron trades. Many women are doing the light kind of drilling, etc. which used to be done by men. We have hundreds of them in Manchester now doing work that was formerly done by men on drilling machines. Women in the iron works were unknown a few years ago, but there are hundreds of thousands of them now."[17]

The Cadbury study of women's work and wages in Birmingham, published in 1907, listed nineteen trades in which women were replacing men.[18] And while Gail Braybon has argued that substituting women for men in jobs that were not deskilled was relatively rare, she does agree that the substitution of unskilled women for skilled men did happen "in some industries when new machinery, requiring less skill on the part of the operative, was introduced, and it was happening in engineering, where unskilled men were also a threat."[19] And she also notes that "they [women] made up part of the labour force in the metal trades, particularly in Birmingham and the Midlands . . ."[20]

The increasing employment of women in the metal industry—particularly the "light" metal trades, such as locks, hinges, toys, nails, jewelry, guns, pots, pans, and other domestic and building ironmongery—was consistent with the trend at Archibald Kenrick and Sons, Ltd. A rare glimpse at the gender division of labor in the cast-iron hollow-ware trade after mechanization and the introduction of female labor can be found in the report of two factory inspectors some years later:

A soft gray iron is used in the manufacture of cast iron hollow-ware. The iron is melted in a cupola and is carried by men or youths in ladles holding about 50 lbs. The pouring of the metal into the molds is also done by them. The method of making the mold varies in the different works. Women, girls, and youths are generally employed on the small sizes while the larger sizes are made by men. In some factories molding machines are used and these machines are often operated by women and girls when the smaller sizes are being made. In some of the older factories the general conditions are poor—roofs are low and the lighting is defective, while in the newer works the foundry conditions are quite good.

When the metal in the mold has set the mold is broken and the casting thrown down to cool. This work is usually done by youths and men. The adhering sand has to be cleaned off the casting. This is generally done by women or youths with wire brushes. This process is, even under the best conditions, a dirty and dusty one. In one works the sand is removed by shaking in revolving barrels. After this cleaning process, the work is examined for flaws. At this stage some hollow-ware is annealed by heating in stoves for about twenty-four hours. Next the fraze or tin left on the casting is ground off at emery wheels and the dust generated by this operation is usually removed by exhaust ventilation. This work is usually done by women, girls, or youths. The inside of the hollow-ware has now to be turned up smooth. In some works this turning is done on automatic machines and the machines are attended by women and girls for the smaller sizes and by youths or men for the larger sizes. Where the turning is done by hand men or youths are only employed. In one works an additional finish is given to the inside by "bobbing" with an emery bob. The dust is removed by exhaust ventilation. In the newer works the conditions in the turning shops are quite good, but in some of the older works the conditions are poor. These latter turning shops are dark, dismal places. The next process is tinning the inside of the hollow-ware. This is always done by a man. The piece of hollow-ware is heated on a hearth. Metallic tin is placed inside and the molten tin is then distributed over the whole of the inside surface with a piece of cork held in tweezers. Sal ammoniac is used as a flux and the fumes are removed by a hood placed over the hearth. The surplus tin is shaken out and the piece cooled by immersing the outside in water. Japanning the outside comes next. This process is entirely carried out by women. The riveting of the handles to the body of the article is the last process of manufacture. Packing is done by men.[21]

Summarizing this trend involving mechanization, and the corresponding employment of unskilled and female workers at the Kenrick firm, Church states that,

The size of the Kenrick's labor force grew slowly until the second half of the nineteenth century, when it rose from about 700 in the seventies to 1,270 in 1894. This increase was accompanied by a trend towards the employment of a higher proportion of females of all ages. This was partly the result of legislation restricting the employment of child labor in factories (which became operative in foundries from 1867), but was also due to mechanisation and the substitution of unskilled for skilled labour.[22]

While the machinery Kenricks installed eliminated the need for skilled craftsmen, these machines could not recruit workers, get them to the factory gates on time, assign them to jobs or machines once inside, make certain they did the work assigned, prevent them from breaking the machinery, or prevent them from wasting raw materials.[23] Nor were the machines capable of inducing the workers to come back to work the next day—perhaps just the opposite. Thus, problems that were once left to the subcontractor defaulted to the directors; and finding solutions would absorb much of their energy in the coming years. That employers felt the weight of this problem is evident in a speech given by A. H. Gibson, one of the Kenricks' colleagues in the Midland Employers' Federation (MEF), before the annual meeting of that organization in 1916:

Of course, any one who knows the outlay of anxiety necessary in controlling a large establishment knows well how much is dependent upon the management, the persons who control it. (Hear, hear.) The workpeople for the most part—I don't say it disrespectfully of them, it is only natural, perhaps—are almost like school children, and unless you have continual methods of control over them, and can stop losses and waste at the inception, you will soon lose all your profits. Every manufacturer knows that is so.[24]

Coming relatively late in the process of mechanization in the Midlands metal trades, this statement shows that, whatever degree of control was inherent in machine work itself, problems requiring "continual methods of control" remained after the subcontractors were gone and the machines installed. What "methods," then, did the Kenricks adopt to achieve "continual control" over their young, largely unskilled, and increasingly female workforce once their subcontractors were gone?

Moreover, who were these workers, where did they come from, and why were they willing to submit to such "continual" methods of factory discipline? In the next section, we address these questions.

The Conditions of the Working Class in West Bromwich, 1900

The success of the Kenrick mechanization strategy depended, first of all, on the availability of young, unskilled, and female labor in the vicinity of the West Bromwich factory. Fredrick Ryland was apparently confident that such labor could be found when he began mechanizing the works in the early 1890s, and his confidence was justified because we have no evidence that the Kenricks had problems attracting such workers at least until World War I.

The short answer to the question of why the Kenricks and other West Bromwich employers had little trouble finding cheap, unskilled labor is, of course, poverty. George Barnsby estimates that 12*s* in 1900 was the "minimum wage necessary to subsist" (i.e., provide food, shelter, and fuel) in the Black Country, while a wage of 24*s* was the "minimum wage necessary to maintain a decent life."[25] And he suggests that "all who regularly received the higher wage were able to keep their families in comfort. These were the aristocrats of labor. All who were unskilled, and most of those in irregular employment, lived at a level insufficient to provide the necessities for a decent life."[26]

Most observers agree that no more than 40 percent, and probably a considerably smaller percentage, of the 10,000 or so working class households in West Bromwich in 1900 were headed by labor "aristocrats," according to Barnsby's criterion.[27] And while artisans were somewhat less likely to send a daughter, or particularly a wife, out to work, even these households were thankful for a few extra shillings per week earned by a young son or older daughter.[28] But for the vast majority of working class households headed by laborers earning an irregular 18*s* or 20*s* per week, the 5*s* to 10*s* from a child or wife was far more a necessity than a luxury.[29]

As discussed earlier, in 1894, 34.4 percent of the Kenrick workforce (24.7 percent male and 9.7 percent female) consisted of young boys and girls between the ages of 13 and 17 (16 for the girls), while another 17.1 percent of the workforce (10.8 percent female and 6.3 percent male) fell between the ages of 16 and 20. Thus, people we would today refer to as "teenagers" made up about one-half of the Kenrick workforce and went a long way toward satisfying the firm's need for cheap, unskilled labor. Sixty percent of these "teenage" workers at Kenricks were male, most of whom were under 17. Sent off to work as soon as they reached the compulsory school-leaving age (10 years of age in 1876, 11 in 1893, 14 in 1899), and sometimes before, these young men took positions in the foundries and shops doing the sort of heavier, common labor boys had always done at Kenricks.[30] What changed for such "lads" toward the new century was that the opportunity to learn a skilled trade, such as turning, via apprenticeship was rapidly vanishing as the Kenricks mechanized and deskilled work at the firm. Thus, where in the past unskilled work might have been a stepping stone to a better paying, skilled job— at least for the luckier boys—after the turn of the century, most faced a lifelong "career" of low-paying, unskilled, or at best, semiskilled, labor.[31]

While the opportunity for a working class boy to become a fully skilled artisan was certainly declining by the turn of the century, over time, young men could move up within the factory hierarchy to better jobs and better pay, and perhaps even into the ranks of foremen and supervisors. For their sisters, on the other hand, opportunities for upward mobility within the factory were almost nonexistent. Cadbury and colleagues reported that in the Birmingham metal trades, employers claimed that the girls "will not learn a trade because, in the first place, they all hope to marry and henceforth to be under no necessity of earning their own living."[32] While no doubt many girls *hoped* to marry their way out of the factory, and some succeeded, it seemed clear to these researchers that the employers were using the girls' hopes to justify their poor treatment of the girls and the profits to be made by keeping them in a low-wage caste. Referring to "press work," a common job at Kenricks, the authors wrote, "In the first place, it is not skilled work, i.e. in a week at the longest a girl has learnt all she will ever learn about her machine. She is taught by the foreman or the next girl, and soon falls into the monotonous regularity of movement which is all that is required for feeding a press. The correct placing of the metal may demand a certain amount of attention, but in many kinds of work it is almost impossible to make a mistake, and girls do not often learn more than they need about the machine itself."[33] In the metal trades factories of Birmingham about the best a girl could do was move a small distance from the dirtiest, dangerous, and worst-paying jobs (e.g., lathe and press work, "dipping," japanning, and "blacking") to the marginally cleaner, safer, and better-paying jobs (e.g., jewelry work, hand lacquering, warehouse work) within, or between, factories.[34]

Finally, in contrast to their brothers, girls who went to work were still responsible, as either wives or daughters, for a "second shift" of domestic chores at home. As suffragette Annie Kennedy observed in 1911:

I saw men, women, boys and girls, all working hard during the day in the same, hot, stifling factories. . . . Then when work was over I noticed that it was the mothers who hurried home, who fetched the children that had been put out to nurse, prepared the tea for the husband, did the cleaning, baking, washing, sewing, and nursing. I noticed that when the husband came home, his day's work was over; he took his tea and then went to join his friends in the club or in the public house, or on the cricket or football field, and I used to ask myself why this was so."[35]

While it is clear that factory girls were disadvantaged in comparison to their brothers, factory work did offer a few opportunities for at least some girls that their mothers would not have had. As evidence from the Cadbury study shows, some young women found factory work liberating and empowering.[36] In the first place, however unrealistic their dreams might have been, most girls wanted and expected to leave the factory upon marriage, and thus tended to see their stint as temporary and, for that, less oppressive than it might otherwise have felt.[37] Second, factory work provided girls who were entering wage labor another alternative to domestic "service" in middle class homes—work which was, until the latter years of the nineteenth century, the most common work available for working class girls. And, according to Cadbury and colleagues, many girls clearly preferred factory work over domestic work, as a few examples of statements written by girls in response to questions submitted by the researchers attest:

> Why I Prefare Working in a Factory Than going To Service—When I was about 14 years of age I went to service for about 18 months and I did not like it at all because you was on from morning to night and you never did know when you were done and you never did get your meals in peace for you are up and down all the time, you only get half a day a week their for you cannot go to Sewing Classes or Christian Indever or any other classes as we do and you never get very large wages in service. And you never know when you are going to get a good place That What I Think About Service.

> Why I Prefare To Work In A Factory—Because there is a fixed time for meals And you know when you are done. you are not all hours of the day And you have only got one to serve and you can go to has many classes has you like in a week you have got Saturday and Sunday to yourself and you can see a bit of life and we are not shut up all day. We have only got one to serve and we have only got one amount of work to do in a day and we can help other girls to go the write way And you can dress how you like in a factory And I Prefare to Work In a Factory. I have got a Sister who is about 15 and she works at the Gold Chain maker And she likes Working in a Factory. There are one or two things which can be approved of The Master ought to have a Lavatory so that you can come out respectable when you were done, there hadent ought to be any bad words to be put out, Most Factories There is, So I remain Yours Truly, F—— J—— .[38]

Additionally, by going off to the factory each day, young girls were also able to escape, for a time, not only household chores, but also the cramped and over-crowded tenements within which most of the West Bromwich working class lived.[39] And what they were also able to escape, much to the alarm and dismay of reformers such as Cadbury, was parental authority. "In a young life just freed from school, free too often from all parental control, surrounded by examples of lives unregulated by any form of restraint, factory discipline is the only controlling influence, and the necessity for promptness, regularity, and obedience may benefit the worker as much as the master."[40] But however successful such factory discipline might have been, and, as we shall discuss below, no matter how much it might have been modeled on the authoritarian family, it was a different kind of authority. Much as Marx suggested, organizing workers under one roof creates the possibilities for worker solidarity and collective resistance to such discipline, and there is evidence running through Cadbury's study and elsewhere that "factory girls" came to see themselves as a class of workers with shared needs, concerns, and interests.[41]

Along with a measure of individual independence and social solidarity, factory work also gave these girls a certain degree of financial independence heretofore unheard of among working class girls. Since the girls were paid directly, and—particularly when on piece-rates—in unpredictable amounts, it was possible for them to hide "pocket money" before the agreed-upon contribution was turned over to their mothers.[42] Thus, the Cadbury study researchers reported on a "social club" consisting of eleven "girls . . . typical of their class, the unskilled factory girl . . ." who in the course of a month together spent 17*s* on such things as "sweets and fruit," weekly visits to the music hall by tram, "music and song books," a "gypsy party" (i.e., an excursion arranged by a group of factory workers), and a weekly subscription to *Evening Home* magazine.[43]

Despite independence, camaraderie, and amusements, work in a factory was usually exhausting, boring, dirty, and quite often dangerous. Moreover, for many women, including those who married unskilled laborers, the unmarried, and the widowed, factory work was a dead-end life of poverty. As noted earlier, there were 354 females working at the Kenrick factory in 1894, of which 250 were under twenty-one, unmarried, and likely living with their parents. Of the 104 adult women workers, only 15 were married, and while it is possible that many of the unmarried women—particularly the younger ones in their early twenties—still lived with their parents, many no doubt were on their own, trying to survive in many cases on less than 10*s* per week. For these women, factory work was, without a doubt, a necessary evil.

To summarize, then, the general level of poverty in West Bromwich, and particularly the dismal state of families headed by laborers, assured the Kenricks a steady supply of cheap wage labor for their factory in the years after 1891. Moreover, not only were such workers unskilled, and to varying degrees dependent upon wage labor, they were, male and female alike, unorganized. And finally, while the state had intervened to regulate working conditions and wages in some trades, and some restrictions on child and female labor were in place in the hollow-ware industry, for the most part the Kenricks had a free hand to exploit the increasingly proletarianized working class of West Bromwich—a class the Kenricks and other capitalists had, in fact, created during the previous century. However, while poverty and dependence might bring a steady supply of cheap and relatively docile labor to the factory gates, the problem of organizing and controlling these workers once they came inside, in the absence of subcontractors, remained.

Really Subordinating Labor

As we have shown, the hierarchy of Kenricks *prior* to the struggles over mechanization in the early 1890s consisted of five tiers. On top, of course, were the Kenricks. Reporting to the directors was the works manager in charge of day-to-day production, who from the early 1880s to 1899 was Fredrick Ryland. Under Ryland's direction were several foremen who supervised, coordinated, and inspected the work of the adult male piece-workers. These men, in turn, employed and directed the labor of several hundred day-rate helpers, usually young men and boys.[44]

Alongside the Kenrick tradition of directly employing apprentices (rather than leaving them entirely to the subcontractors), the Kenricks also hired foremen to supervise the subcontractors, so the subcontractors had somewhat less autonomy than was typical of other firms in the metal trades.[45] Moreover, the existence of foremen at the firm, prior to the elimination of the subcontractors, suggests that the transition from subcontracting to a more complex hierarchy of labor control might have been somewhat smoother for the Kenricks than it often was for other metal trades employers.[46]

But despite the existence of foremen, the firm's subcontractors had been central to the labor process. They not only brought their skill to the factory, but they also brought the helpers whose labor power they exploited and behavior they policed, leaving worries about capital, technology, and marketing to the masters. It is no surprise, therefore, that when the Kenricks decided to eliminate these pivotal workers, beginning in the 1890s, authority relations at the firm changed. One way the Kenricks seem to have dealt with the supervisory vacuum was to simply increase the number of foremen at the factory. Thus, even though the Kenricks had only "several" foremen to supervise and inspect the work of the subcontractors in the 1880s, by the early years of the twentieth century there were possibly as many as forty.[47]

Littler suggests that in small- and medium-sized firms lacking a systematic, or "scientific management" ideology, foremen tended to have the same authority as the subcontractors they had replaced.[48] As described by one employer:

> In most works . . . the whole industrial life of a workman is in the hands of his foreman. The foreman chooses him from among the applicants at the works gate; often he settles what wages he shall get; no advance of wage or promotion is possible except on his initiative; he often sets the piece-price and has the power to cut it when he wishes; and . . . he almost always has unrestricted power of discharge.[49]

We cannot say for certain that this description fits precisely the Kenrick foremen, but in a firm with approximately one thousand workers and, at most, a half a dozen directing partners, authority over such day-to-day matters as these probably was delegated to subordinates. Furthermore, the Kenricks were widely known for devoting time to civic matters that often would take them away from the factory. And finally, in the minutes of the directors' monthly meetings of this period, we find almost no discussion of "personnel" matters concerning any workers other than foremen and office staff; indeed, unskilled and semiskilled production workers were rarely mentioned in these proceedings. The firm's few remaining skilled workers were mentioned, as a group, usually in connection with changes in wages. As we will discuss below, it is only when the Kenricks were forced to bargain collectively with their unskilled workers, as a result of the strike in April of 1913, that we see the wage rates and other matters concerning the "bottom dog" becoming a matter for discussion at the directors' meetings.[50]

Although we cannot speak with certainty about the breadth of the foremen's authority, evidence from accounts of two strikes that occurred at the firm in 1913, one in April and one in October (more on these strikes later), both of which were prompted by conflict between workers and foremen, suggests that these men had considerable power and autonomy in dealing with their subordinates. In the first instance, a foreman apparently had the authority to dismiss a request from employees concerning shift work. His action was fully supported by the directors of the firm and this lead to a work action.[51] In the second instance in October, a dispute arose over a shop foreman whose "dictatorial and bullying behavior" was resented by female subordinates. As in the earlier case, the Kenricks backed their foreman, suggesting an unwillingness to undermine his authority despite the accusations leveled against him. The evidence should not be taken to imply that all Kenrick foremen were abusive, but it does clearly indicate that they were in a position to act "dictatorial"—with the backing of their superiors—if and when they chose to do so.

Piecework for the Unskilled

Earlier generations of unskilled workers at Kenricks were paid day wages, but with the decline of subcontracting, many unskilled and semiskilled workers were put on piecework. As Littler points out, cutting piece-rates was a strategy used to reduce wages or, by paying a "premium bonus," a common means of speed-up.[52] As we will discuss in connection with the strike in April of 1913, low piece-rates were a central complaint of the strikers, particularly the female workers, suggesting that the Kenricks, too, may have used piece-rate cutting as a way of intensifying effort. But since there is almost no discussion of piece-rates for unskilled workers in the Directors' Minutes before 1913, these rates were probably set by foremen or shop managers.

In addition to, or in support of, the use of foremen and piece-rates to control their growing and unskilled workforce of "schoolchildren," the Kenricks wrote, and presumably distributed to all workers, a document entitled "Rules to be Observed in the Works of Messrs. Archibald Kenrick & Sons." This document, which we date circa 1900, not only reflects the Kenricks' disciplinary regime, but also gives a feeling for what it might have been like to work at the foundry at this time:

1. The hours of Work on all days, except Mondays and Saturdays, are from 6 o'clock till 6 o'clock in the Evening; on Mondays from 7am till 5 pm, and on Saturdays from 6am till one pm. Half an hour is allowed for Breakfast, from half past 8 to 9am, and hour for dinner from one to two pm, during which time the engines are stopped.
2. All Work-people must be in by the hour appointed for commencing work. The Bell is rung 5 min. before that time. Any person entering between 6 and 6-30 loses half an hour, No person is admitted between the hours of 6-30 and 8-30; 9 and 1; or after 2. No Workpeople are allowed to leave during Working hours without a signed check.
3. The hours of the Warehousemen and Girls, Packers and Odd work Blackers, (except the Scourers, who work according to Rule 1) are from 7 o'clock in the morning until 6 o'clock in the Evening; leaving off at 1 o'clock on Saturday.
4. Overtime is Reckoned at 2 hours per Quarter.
5. Any workman who does not keep the above named hours is liable to be fined at the discretion of the Managing Partners.
6. Every person Employed in the Works has a check given him, which must be deposited in the porter's lodge on Entering. These checks are returned at 6 o'clock, when work is done. Any person omitting to deliver up the check on entering or to receive it before leaving loses the time during which they neglected to pay attention to this rule.
7. Every person must deliver up his or her own check, and anyone asking for anothers is liable to instant dismissal.
8. Every person losing his or her check must pay 3*d* for a new one.
9. All persons having their meals sent to the Works must apply for them at the Lodge after the Bell has been rung. No one is allowed to enter the Works who is not employed therein.
10. Smoking is not allowed during working hours. Any persons Smoking in the Warehouse, Pattern Shops, Carpenter's Shops, or Risk Yard will be fined 2*s*-6*d*, in other jobs 1*d*.
11. In all cases of leaving employment, a fortnight's notice is given and required.
12. No dogs admitted.
    N, B, The books are made up to Thursday night, and Wages paid at 1 o'clock on Saturdays.[53]

Perhaps it should not be surprising that among employers who thought of their workers as "schoolchildren," this system, with its fairly detailed specification of deductions, fines, threats and punishments, looks very much like what we would expect to find in place at an authoritarian day school. Clearly, it would have taken a company of diligent foremen and clerks to implement and enforce these rules—and undoubtedly many additional informal ones—in a sprawling works of over 1,000 workers.[54]

Some Despotism, Some Paternalism, Some Sexism

The authority structure that emerges at Kenricks in the years between 1892 and 1913 suggests a factory despotism—or at least authoritarianism—with plenty of rules and regulations, rewards (however meager) and punishments, along with a large supervisory staff to implement them. This, however, is not the whole picture. The evidence suggests that between the decline of skilled labor in the 1890s and emergence of collective bargaining for unskilled workers in 1913, the Kenricks mixed elements of both paternalism and despotism as the situation and circumstances seemed to require. In this sense, they were much less managerial ideologues than they were problem-solving pragmatists.

The authoritarian side of the firm, evident in our discussion above, is consistent with, and no doubt to some extent flowed from, George Kenrick, the man who held the firm's top post for some forty years (1898-1938). George Kenrick (1850-1939)—Sir George after 1910—joined the board of directors in 1873 after attending university and apprenticing as a mechanical engineer at the nearby Smethwick workshops of Nettlefold. First cousin to John Arthur (1829-1926), the founder's grandson, George assumed the chairmanship of the board when his cousin stepped down in 1898.

Based primarily on interviews conducted in the 1960s with knowledgeable family members, Church paints the following portrait of Sir George:

> Throughout the early decades of the twentieth century until his voluntary but long-awaited retirement in 1934, Sir George remained at the apex of the hierarchical managerial structure, ruling the company in a thoroughly autocratic fashion. The lines of communication to the chairman, who remained aloof from his subordinates, did not permit the movement of ideas and even Byng (Kenrick, Sir George's successor as Chairman), whom Sir George patronised as the company's crown prince, for a long time deferred to the views of his elder cousin whenever a difference of opinion arose. Not until some time in the twenties did Byng begin discreetly to countermand some of his more preposterous instructions, and on occasion, after lengthy argument with Byng, Sir George would even admit to having been in error.

> it seems that—especially in the traditional lines—Sir George took the view that Kenrick's products were unequaled, and that price cuts were symptomatic of weakness. Moreover, his apparent disregard for turnover, as reflected in his readiness to raise the price of an article as soon as it began to sell well, was particularly frustrating for his salesman and managers (none of whom possessed authority to make quotations under any circumstances) . . . Such a sales policy, which has been described by the manager of the London office as "rigid and unimaginative" has been attributed to his lack of contact with customers. Coupled with his apparent insensitivity to customers' wishes, Sir George's unapproachability helps to explain why the suggestions advanced by his salesman, not to mention the company's customers, received scant attention. A typical remark of Sir George,

according to the head of an Anglo Department in the general office, was "You are not paid to think, we do the thinking and you do as you are told."[55]

If, in Sir George's view, an upper-level manager of the firm was "not paid to think" he probably expected considerably less than thinking from the hundreds of teenage "factory girls" his managers hired to operate punch presses and paint bolts for 8 or 10 shillings a week. And it is doubtful that his knighthood, conferred during his reign as "Lord Mayor"of Birmingham, brought him any closer to the working people inside his firm or in the community.[56] Indeed, if customers received "scant attention" from Sir George, it is doubtful he paid his workers any attention at all—at least until they demanded it.

But, as was discussed earlier, the Kenricks also had a long tradition of paternalism, and while elements of this tradition did make its way into the twentieth century, it generally did not make its way down to the shop floor. As Church reports, foremen and clerks received favored treatment, not only in terms of pay, but also with respect to pensions, regular wage increases during good times, and some measure of job security.[57] In contrast, at no time, up until the government introduced its old age superannuating scheme in 1908, did the Kenricks offer pensions to ordinary workers in other than an ad hoc fashion, and only then after a lifetime of service to the firm; there was no routine mechanism for sharing the firm's good fortune with the workforce; and workers were subject to unpredictable reductions in hours, or layoffs during slack periods.[58]

Thus, with respect to the most substantive elements of paternalism, the Kenricks seem to have adopted a "dualistic" approach where, as Littler points out, the firm would, rather than "casualize" all labor, retain and treat well a core group of valued employees while leaving the mass to "float."[59] While core employees received decent pay with raises, job security, and a pension, the mass of unskilled workers were offered, in times of sickness, two nurses (1891); land for cricket and football (1907); and a works recreation club (1908).[60] These offerings, provided exclusively to Kenrick employees, were dwarfed by the Kenricks' extensive involvement in, and monetary contributions to, civic and community institutions such as the West Bromwich hospital, the General Hospital in Birmingham, as well as the many contributions the firm made to the founding and funding of schools in the district.[61] In this way, the Kenricks believed they were "giving back" something to the community, and, indeed, they were, but at the same time they were, as we have seen, rather miserly with respect to specific welfare provisions to their own workers. No doubt if asked whether they would have preferred pensions, secure employment, and decent wages over corporate philanthropy, the bulk of the firm's unskilled workers, most of whom, as we have seen, lived in poverty, would have likely preferred the former. But distributing the firm's wealth in the form of wages and benefits to workers rather than giving some of it away ad hoc to community institutions would have neither provided a secure position within the upper-middle class, nor guaranteed that the wealth would be used to further those interests. Thus

Kenrick paternalism, other problems aside, contained a certain amount of hypocrisy. As Richard Trainor reports in his study of Black County elites, "local radicals pointed out that families such as the Bagnalls cut wages as vigorously as they provided schools and chapels," and one might even suggest that they did the latter in part to compensate for and cover the former.[62]

A final element in the complex way in which the Kenricks treated their workforce, in the years between 1892 and 1913, concerns the sex-caste system they installed within the factory. This sexism worked not only to increase profits, but also to enhance the authority relations at the firm. In what sense was the Kenrick factory organized as a sex-caste system? First, as Ryland discovered in 1891, unskilled girls and women could be paid less than similarly unskilled boys and men simply because they were female. Thus, the Kenricks were quite willing, it appears, to incorporate beliefs about female inferiority into their factory. Second, females were restricted to only the lowest ranking and most poorly paid jobs. And finally, females who did work at the firm were never put in a position of authority; conversely, they were always supervised by men, a situation that merged the cultural authority of male over female with the organizational authority of supervisor over subordinate.

Aside from these institutional arrangements, we can also find more subtle hints of the sex-caste system at Kenricks. In particular, the almost complete absence of reference to production workers, especially to women workers, in the written records of the firm also reveals, we think, a lack of respect on the part of the firm's directors toward these women. This is particularly evident when compared to the attention paid to male employees. For example, woven into the record of technological change at the firm reported earlier, we also find the following references to the passing of the generation of men who had worked at Kenricks since, in some extreme cases, the early years of the nineteenth century:

> May 15, 1895 "Letter from W. H. Cooper, foreman carter, acknowledging gift of silver tea service on retirement after 40 years with the company."
> February 19, 1899 "Death of Fred Ryland reported: 32 years associated with the business."
> March 21, 1900 "Reported death of A. Smith, foreman in Axle Pulley Shop" ("Suttons").
> October 21, 1903 "'In recognition of his 60 years of service' a silver teapot presented to Tom Stephens, blacksmith."
> December 16, 1903 "Scheme for retirement of heads of departments all approaching 65 years produced by G.H.K." viz.:

Arthur Underhill (63)	to retire in	1905
William Mould (58)	to retire in	1911
Sam Withers (63)	to retire in	1904
John Sample (62)	to retire in	at 65
John Warrington (61)	to retire in	at 65
Joseph Woodhall (62)	to retire in	at 65

May 17, 1905 "J.A.K. reports death of Woodhall retired foreman carpenter and T. Holden retired tinner."
December 20, 1905 "Reported Arthur Underhill's decease; allowance of 2 pounds per week to the widow for one year granted."
August 14, 1907 "John Evans who had completed 65 years in the employment of the Company and its predecessors retires on pension of 10/- per week."
October 16, 1907 "John Sample, foreman engineer, retires."
May 20, 1908 "J. Howle, senior, retires on pension 10/- per week. J. Gough Baldwin's retiring allowance agreed at 175 pounds p.a. Jack Warrington, Foundry Foreman, to retire next month after 57 years of service."
July 15, 1908 "Inscribed Gold Watch presented to J. Warrington in June. Philip Chumbley, senior, head warehouseman, to retire in August. J. Morse to succeed him."
June 15, 1911 "W. Mould retires and Walter Gale succeeds him."[63]

There are, of course, practical reasons why the directors had to deal with such matters at their meetings. But however much these deaths and retirements were noteworthy for practical reasons, we think their symbolic importance is even more significant. What we have here are the directors of the firm honoring, remembering, and rewarding the *men* who for years had made Kenrick hollow-ware a household name. It is significant that, nowhere in the Directors' Minutes, do we find any similar reference or recognition of *any* women employees at any time during the firm's recorded history. When women are mentioned—as in 12-20-05 above, where Arthur Underhill's widow is ordered to receive a small pension—it is through their connection to men, usually loyal male employees. Neither the women who worked for the Kenricks, nor the women married to the men who worked for the Kenricks (not to mention the Kenrick women themselves) are ever recognized or honored in their own right.

Paternalism and Proletarians

Between 1891 and 1913, the firm of Archibald Kenrick and Sons Ltd. produced and sold more hollow-ware, hired more workers, and earned more profits (over £600,000 in dividends were paid to shareholders) than at any time in the firm's history. While not the only factor responsible for the success of the firm during this period, it was no coincidence that this was also a time during which the firm shifted from a relatively skilled, expensive, organized, and powerful workforce toward a relatively unskilled, unorganized, and powerless workforce.[64]

When it came to the problem of organizing and managing their changing workforce, the Kenricks met some familiar and new challenges with a combination of familiar and new tactics. While continuing and extending the firm's tradition of a "stern yet considerate paternalism" to all workers at a symbolic level, the directors were careful to reserve its most substantive elements, such as pensions, steady employment, and good wages, for a core of valued clerks, managers, and foremen.[65] In contrast, the increasing number of younger, unskilled, and female workers were paid poorly and treated as a casual labor pool of unruly children to be disciplined, or replaced, as required. Running through and enhancing this disciplinary regime was a sexism in which female workers were treated as inferior to male workers. From the start, the Kenricks sought to hire female workers because they were less expensive than equally unskilled male workers; and once hired, women were ghettoized into the most poorly paid jobs where they were subjected to the combined managerial and cultural authority of male foremen and supervisors.

Clearly, the Kenricks had an interest in maintaining and perpetuating the dominant and profitable position they had struggled for and achieved by the turn of the century. But, of course, the Kenricks were not the only ones engaged in this struggle. Other capitalists throughout the Midlands metal trades and across Great Britain were trying to achieve the same goal, through largely the same means. As a result, the size of the proletarian class was growing; and, throughout the metal trades and across Britain, unskilled wage-workers, including women, were, for the first time in significant numbers, beginning to organize themselves to resist their exploitation. As a result, between 1910 and 1914, English workers battered English capitalists with wave after wave of strikes. Summarizing the period, Hugh Clegg writes:

> The annual number of stoppages recorded by the Labour Department of the Board of Trade, which had not exceeded six hundred since 1901, climbed through 872 in 1911 to a peak of 1,459 in 1913. Except in 1908 and 1910 the number of working days lost through stoppages had not risen above four million since 1901, but from 1911 to 1914 it averaged nearly eighteen million, and in 1913 reached the record figure of over forty million.[66]

Historians suggest that the most immediate cause of the "great unrest" of 1910-1914 was worker discontent over the steady decline in real wages that began in the 1890s.[67] Between 1875 and 1895 the purchasing power of English working class families increased by approximately 40 percent as a result of the deflation that occurred in the 1870s. In the next twenty years, however, between 1895 and the outbreak of the First World War, stagnant wages and increasing prices dramatically eroded working class living standards, with the average working class family spending more of its income on food in 1914 than it had in 1900.[68]

Working class anger might not have exploded into what Eric Hobsbawm calls the "mass militancy" of 1910-1914 if the hard times had been shared by all English families. But the suffering was not evenly distributed; indeed, the upper class did not suffer at all.[69] Even Arthur Taylor, in what might otherwise be read as an apology for Edwardian excess, acknowledges that in the four years between 1910 and 1914 the share of national income going to the "profit-earning classes" increased by more than 15 percent.[70] Moreover, according to Hobsbawm, the working class was becoming more aware of this increased inequality:

> the working class was not so much segregated as alienated from the ruling class by two developments which, together with the fall in real wages, Askwith [the government's chief labour negotiator at this time—auth.] made responsible for the labour unrest of 1910-1914. These, he told the Cabinet confidentially, were the conspicuous display of luxury by the rich, especially demonstrated by the use of the motor-car, and the growth of the mass media, which made for a national coordination of news- and activity . . . the wealth of the rich was now more visible and more resented.[71]

Along the same lines, George Dangerfield, in his lively account of the "death of Liberal England," reminds us that the revolt may have resulted more from the way capitalists treated their workers than from the way they spent their money:

> The instinct of the British worker was very active in 1910. It warned him that he was underpaid, that Parliament—left to itself—would keep him underpaid; it told him that good behavior had ceased to have any meaning; it asserted that he must unite at all costs . . . The first steps into the Unrest seem straightforward enough—anger at the fall of real wages, at capitalist aggression, at the unwillingness of Parliament; anger fomented by agitators, and informed by vague fears, and leading to solidarity.[72]

And the view from the vantage point of capital:

> A capitalist might understand, and even condone, the concentration against him of foreign capital, but under the circumstances he could only feel extremely uneasy about any sign of agitation among the workers, from whose labours, between 1900 and 1910, he had realized a considerable profit. By going slow on wages, he could store up something against a rainy day . . . As for the workers, he did not expect them to see eye to eye with him, and the only method that suggested itself was to give them a black eye the moment they showed any disposition to see at all. This is what was done in the early years of the twentieth century—and as a feat it was generally applauded by the middle classes who—themselves deprived of economic power and reduced to a mere assortment of clerks, salesmen, officials and civil servants—looked upon the producers with a jaundiced, a fearful, a vindictive gaze.[73]

Moreover, as social reformers documented and criticized the growing gap between rich and poor and the dismal state of much of the working class, and governments grew nervous about the potential civil disorder of industrial conflict and unrest, the British state began taking a somewhat closer look at the costs, as well as the benefits, of its long-standing reluctance to interfere with the "autonomous workings of industry." In combination with these broad, societal trends—

most of which the Kenricks had no control over—the Kenricks themselves set the stage for rebellion among their unskilled workers, and particularly their female workers, by treating them so poorly. By refusing to extend their paternalism to all of their workers, the female workers in particular, the Kenricks left themselves open to the possibility of having these workers organize against them. This possibility was realized on the morning of 7 April 1913 when the Workers' Union led unskilled workers at Kenricks and other factories on Spon Lane in a two-week strike for union recognition and a minimum wage.

7 April 1913: Walkout at United Hinges, Ltd.

Existing accounts agree that the strike began with a dispute in one shop between unskilled male workers and their manager that escalated into a standoff when the Kenricks refused to deal with the Workers' Union.[74] Church's summary of the trouble is a useful place to begin because it is based in part on the testimony of the worker, Jack Jones, who claims to have started it:

> Sometime after the installation of the cold rolling mill at United Hinges, the eighteen men affected had requested to be allowed to work alternate nightly and daily shifts week by week and to be paid a regular 21s weekly, instead of 26s after the "nights" week and only 16s after the "days" shifts. Their spokesman was Jack Jones, then aged twenty-five and unmarried, who had only recently joined United Hinges. It was he who gave voice to the men's fears that a week's pay of 16s would bring hardship to their families. When Bradley, the shop manager, dismissed the men's request, Jones and the men resolved to involve officials of the Workers' Union and to push for the Union minimum. Faced with formal representation, John Archibald (Kenrick) declared that it had not been the firm's policy to recognise the trade unions, nor would it be in the future. Refusal to negotiate led to the strike.[75]

The Workers' Union (WU) was one of the "new unions" founded in the 1890s to organize the industrial proletariat that the Kenricks and other employers were creating as they mechanized and deskilled production.[76] While open to all non-unionized workers, male or female, by 1910 the WU had managed to enroll only 5,000 members in 111 branches. The WU hit the Kenrick firm about the time it was hitting its stride. Between 1910 and 1914, the WU saw such growth that by the latter year it had 91,000 members in 567 branches, with most of this growth coming in the tumultuous year of 1913.

In the metal trades industries in and around Birmingham and the Black Country, the WU focused its efforts on forcing employer recognition and then fighting for a scale of minimum wages. The union's approach was to first agitate at a factory among the unskilled males for a 23s minimum, and then, when they felt they could win a confrontation, present demands to the management.[77] As Church's previous account shows, the WU had already gotten to Jack Jones, who told Church in the 1960s that at the time of the strike he thought of himself as "a self-appointed unpaid organizer and delegate for the Workers' Union," suggesting that the WU was behind the walkout on the morning of 7 April.[78]

Undoubtedly, the Kenricks were aware, in the weeks and months prior to the strike, that the WU was agitating among their workers. And they had good reasons to fear an organizing drive at their firm, because the WU had, in recent months, led unskilled workers in highly visible and successful strikes against two nearby firms: the weighing-machine firm of W. & T. Avery in November 1912 and the engineering firm of Tangyes in February 1913.[79] And while we do not know whether, prior to 7 April, the Kenricks made any efforts to forestall the WU by encouraging their workers' loyalty to the firm, we do know their paternalistic appeals and laments subsequently fell on deaf ears.

We also do not know what John Archibald Kenrick *expected* Jack Jones and his colleagues to do when the owners initially refused to let the WU represent them, but a portion of the statement by his uncle, Sir George, issued the following Saturday (*The Free Press*, 15 April 1913), suggests that John Archibald and his codirectors *wanted* the workers either to get back to work or to hash out the problem with the owners, without the union's interference:

> The management claim that they are paying, and have been paying for some time, previous to recent advances, the present full standard rate of wages for laborers and female labor in the district. The questions at issue have been discussed with deputations of our workers, and the few cases on individual hardship which are being dealt with might have been amicably settled in the ordinary way.

The workers' unwillingness to settle their differences with the Kenricks either "amicably" or "in the ordinary way" represented an obvious threat not only to profits, but also to the Kenricks' paternalist tradition. Thus, in John Kenrick's refusal to negotiate, we find the "father" attempting to reassert his right to deal with his "children" as he alone sees fit.

The "father" was also apparently hurt and angered by the "children's" misbehavior. Thus, when the press tried to get the owner's side of the story on Monday afternoon, they could only report that:

> At present it is impossible to give employer's version of the matters in dispute, because inquiries at both works only elicited the reply that they are themselves in ignorance of the causes of the trouble . . . It is stated that they have not received any complaints from the workpeople, who absented themselves without preliminary notice of any kind, and without any intimation that they had grievances which they desired redressed. Under these circumstances the employers state that they have nothing to communicate.[80]

If Jack Jones's testimony about the argument in the cold rolling mill is true, the Kenricks knew much more, even on Monday, about what their workers wanted than they were willing to let on, but posturing in this way worked to position the strikers as disobedient and irresponsible children. This point is made even more forcefully in the statement the Kenricks released days later, having finally formulated their version of the events:

> On Monday morning one of the helpers in the rolling department at United Hinges Ltd., who had been working days, and who was asked to go on the night turn, objected. He saw his foreman during his breakfast half-hour. The foreman discussed the matter for some time. It being breakfast time, the man suggested that he should come again after he had his breakfast, at nine o'clock. The helpers then left the works, and instead of returning as proposed, persuaded all the other workers in their department to leave also. The employees at Archibald Kenrick and Sons Ltd., went out in sympathy, without knowing what the grievance was. From Monday at 9 a.m. until Wednesday morning no information of any kind was communicated to the management of the United Hinges as to what the trouble was, and it was not until noon on Friday, four days after the employees of Messrs. Archibald Kenrick and Sons Ltd. had left their work without notice, that the secretary of the Workers' Union brought the details of their complaints.

> The mayor (Sir George Kenrick, then Chairman) also complained that the men went out on strike before consulting the Union at all, and he considered that they had acted very foolishly in doing so. But after they had placed themselves in an awkward position they sent for the officials of the Union. His worship contended that Messrs. Kenrick and Sons were actually paying more than the rate of wages current in the district, and he suggested that the Union should turn their attention to other firms, who are paying less, and bring them up to that rate, and then they could ask his firm to consider the question of a further advance.[81]

Given all the Kenricks felt they had done for their workers and their community (feelings that were not without foundation), it is not surprising that they would react so strongly to their workers' disloyalty. And an event that occurred on Monday afternoon probably added a good deal of insult to the injury. After their arrival in West Bromwich on Monday afternoon, WU organizers George Geobey and Julia Varley assembled the strikers for a rally on some "waste land" nearby.[82] But as James Leask and Philomena Bellars report in their account of the strike, the land used for the rally was, ironically, supposed to be a public park that had been donated to the town of Smethwick years earlier by the Kenrick family.[83]

After offering to meet with the owners, strikers assembling outside United Hinges on Tuesday morning found the following notice "affixed to the gateway of their works":

> The Directors of Archibald Kenrick and Sons, Ltd., will be glad to meet a deputation of their employees not exceeding six in number to hear the reason that has

caused them to cease work without notice. The deputation can be received at noon today.[84]

The workers didn't show up at noon because they were insistent that they be represented by the WU, and the union representatives were already busy, Tuesday morning, in discussion with the owners of Guest, Keen, & Nettlefold, on behalf of the workers who had initially come out in support of the United Hinges workers, but who had now formulated their own demands. But the Kenrick workers did meet later that day with the owners, and Jack Beard, Secretary of the WU, was there to represent them. Thus, while the Kenricks continued to talk tough throughout the first week of the strike, the evidence shows that they were already meeting with representatives of the WU, who were acting on behalf of their workers, as early as Tuesday afternoon. The willingness of their neighbors at Guest, Keen, & Nettlefold to bargain with the WU, a severe blow to employer solidarity on the block, surely disheartened the Kenricks and undoubtedly contributed to their willingness to deal with the union. But what probably got them to the bargaining table in a hurry was the news that unskilled Kenrick workers "on the old side" in the hollow-ware works had, on Tuesday, left their shops and had joined the workers from United Hinges.

The early meetings between the two sides seem to have produced very little progress and it was not until Friday, 11 April 1913, that the *Birmingham Gazette* was able to report that "negotiations for the settlement of the strikes in progress at West Bromwich were proceeding satisfactorily." Progress was, however, uneven. While the talks were apparently going well at Guest, Keen, & Nettlefold, and also at United Hinges— about which it was reported, "The firms have shown themselves quite willing to discuss matters with the representatives appointed by the employees . . ."—on the "old side," at the hollow-ware works:

> it has been found impossible by the Union officials to take any definite steps toward a settlement. The position is complicated by the fact that additional sections of the employees are joining in the strike, those engaged in the clay and enamelling departments ceasing work yesterday. This involves some delay in preparing the case for the workpeople.

But more ominous, perhaps, than the complication of having to deal with the grievances of additional strikers was the news that:

> The strike has already rendered the larger part of Kenrick's extensive works idle. In addition to the hands who have left their work voluntarily, a number have had to stop because of a lack of material, due to the suspension of operations in other departments. The majority of the firm's employees have now ceased work.

On Monday, 14 April, the one-week anniversary of the strike, things started to go bad for the Kenricks. First, news came that the strikers at Guest, Keen, &

Nettlefold had gone back to work "upon terms which embody practically the whole of their demands" creating "much satisfaction both among the workpeople directly concerned and also those engaged in the other disputes."

In addition, they also learned that the disputes at the two Kenrick firms had "assumed a rather serious aspect." The aspect in question was a near riot that occurred outside the hollow-ware works Monday afternoon when seven to eight hundred "strikers and their sympathizers" blocked the gates in front of the Kenrick works, and, despite the presence of police who were unable to clear passage, prevented a much smaller number of Kenrick workers who had remained on the job from returning to work after lunch. That evening the Kenricks, perhaps fearing another "scene of disorder," announced that the works would be completely closed the next day.

The Kenricks' sense of isolation was deepened by the news that other firms in the borough were falling like dominos to the WU. Attempting to avoid trouble before it started, Chance Bros. and Co., of the Spon Lane Glass Works, Grigg and Sons, hauliers, Spon Lane, and the Muntz's Metal Company, of Smethwick, all announced they would pay the 23s minimum for "general laborers," that is, men. And then, on Tuesday, all of the workers at W. Cross and Sons, iron founders, West Bromwich, ceased work and demanded an increase in wages. The strike was reported to be "brief, decisive, and quite successful from the worker's point of view" as the employers "promptly acceded to the striker's demands, and in a remarkably short space of time the hands had returned to their duties and work was proceeding as usual."

Furthermore, while the Kenricks, in their statements to the press, were clearly playing to public opinion, the WU was playing the same game, but winning it. Because so few of the workers at the affected firms had been with the WU more than six months (most had not been with the union six days), almost all were ineligible for strike benefits. Thus, money had to be raised to feed these workers and their families since their cupboards would be bare within a week. And so "local collections were organized and barrel organs toured the streets of Smethwick and West Bromwich, collecting on behalf of the strikers."[85] And "During the first week £60 were collected, and on Sunday, 13 April, a parade was organized in West Bromwich, and 3,000 people took part."[86] The success of the parades, collections, and rallies held by strikers and their sympathizers seemed to both reflect and shape public opinion in favor of the strikers and against the Kenricks. And, of course, as prominent citizens, elected officials, and erstwhile "new model employers," the Kenricks were particularly sensitive to public opinion.

But while the struggle seemed to be tilting decisively toward the WU, people were evidently uncertain how the Kenricks would respond. The *Gazette* coverage of the strike for 16 April 1913 followed a report on the quick strike at W. Cross and Sons with an update on the dispute at Kenricks:

The action of the Kenricks in closing down the whole of their works, with the exception of the offices, prevented any further development of the disorderly scenes which were witnessed the previous day in the vicinity of the premises, when the hands who were remaining at work suffered molestation from the strikers and their friends.

In some quarters the decision of the firm to close the works has been interpreted as indicating the intention of contesting the issue with the disaffected employees, in which case it is not difficult to foresee troublous times, because the strikers have adopted a very determined attitude. They maintain that their demands are reasonable, and they point to the fact that they have been conceded by other firms in the district as proving this.

It was announced yesterday that the number of employees of Messrs. Kenricks and United Hinges who are on strike is 1,122, this being the number enrolled as members of the Workers' Union.

At the same time, the WU reported that there was no progress in their talks with the Kenricks, and, citing data from "wage sheets" they obtained from the workers, rejected Kenrick's claim that the firm was already paying the 23s minimum to unskilled men and "district rates" to skilled workers.

But "troublous" times were not ahead, as it turned out, because the Kenricks apparently had no stomach for the fight. On Friday, 18 April 1913, the *Birmingham Gazette* reported that a "long conference" had taken place the previous day between the two sides "at which the demands of the employees were fully discussed." And at a rally held after the meeting, Jack Beard, the WU representative, told the strikers that while he had nothing definite to report at the moment, he felt that the owners were coming around to their position, which, in fact, they did the next day, when Beard announced to the strikers that he had "every reason to believe a satisfactory settlement will be arrived at."

The centerpiece of the agreement that brought the strikers back to work at 9:00 A.M. Monday morning, 21 April, two weeks to the day after the strike began, conceded the 23s minimum for adult male workers and a 12s minimum for adult female workers, with proportionately lower wages for younger workers, male and female. In addition, the Kenricks also agreed that the men in the cold-rolling mill would receive a standard 23s per week regardless of what shift they worked; that they would end the practice of withholding payment from the female pieceworkers until their work was shipped from the factory; and that a slightly higher rate would be paid for girls who worked in the plating, enamelling, and blacking departments, where the acids destroyed the workers' clothing.

Most observers viewed the outcome of the two-week strike at Kenricks as a major victory for the workers and a clear defeat for the employers. The Kenricks themselves indicated as much when they noted in the firm's Annual Report for the year ending 30 June 1913 that:

During the Spring, a strike without notice broke out in a small section of the workers and fanned from the outside gradually extended so far as to cause the complete stoppage of the works for a short time. Prolonged negotiations led to a settlement by which many workers employed at the lower rates of wages have received substantial advances. This and the full brunt of the Insurance Acts have added to the general costs.[87]

There is also additional support for the conventional view that the strike was "a victory for the workspeople"; most obviously in the same ways it was a defeat for the Kenricks.[88] Leask and Bellars, in their celebratory account of the strikes that summer, note how different workers might have found satisfaction in the outcome:

The demand for a minimum wage was an accurate reflection of the need and desire of the average worker. In the minds of many it had become more than just a figure of 23*s*. It was their belief that if there could be established in industry a minimum wage below which no man could earn less, then this was the beginning of industrial security.

To other workers the wage demand was the birth of a new virile movement attempting to establish itself . . . To those of hard-headed practicability it represented a substantial improvement in the existing rates of pay, and brought with it a promise of a higher standard of living.[89]

While an improvement in wages and the realization of "industrial security" via the minimum wage were important achievements, the most important victory for the Kenrick workers was, in our view, forcing the owners to recognize and deal with the Workers' Union. With this declaration of independence, the workers turned their backs on the Kenricks and brought an end to years of paternalism. And while they clearly did not cast off the chains of wage labor, they certainly managed to loosen them a little.

But What about the Women?

As the percentage of female workers at the firm increased, beginning with the struggles over mechanization in the 1890s, profits at the Kenricks and other Midlands metal trades firms became increasingly dependent on their continued ability to exploit this pool of female workers. But as the above account of the strike indicates, the WU did not challenge the practice of paying "unskilled" girls and women less than unskilled boys and men of the same age; instead, they mobilized male and female workers to press for a *scale* of minimum wages differentiated by sex and age—with younger and female workers receiving less.

But however reticent the WU was to challenge male privilege, there is evidence that at least a few of the workers, female and male, were aware of this inequality, thought it unjust, and were prepared to challenge it. Indeed, the workers involved in the walkout expressed such an interest, among other concerns, on the first day of the strike. On 8 April 1913, *The Birmingham Gazette* began its coverage of the strike with headlines about a "labor dispute" revolving around "women workers":

> For some time past there have been symptoms of unrest among the women and girl workers of West Bromwich, who form a very numerous section of the local industrial community, and the West Bromwich Trades Council have gone so far as to appoint a committee to investigate the wages and general conditions of female labor in the borough.

> Yesterday the smouldering fires of discontent burst forth into flames. Upon their own initiative over 100 men and young women in the employ of United Hinges, Ltd., of Spon Lane, West Bromwich, came out on strike. In the afternoon they were joined by about 120 male and female employees of Messrs. Guest, Keen, & Nettlefold, of the Stour Valley Works, Spon Lane.

> The strikers are demanding, and a minimum wage of 10s per week for all adult female workers . . . Most of them are members of the Workers' Union, and they telegraphed to the headquarters in Birmingham that they had come out on strike. Mr. G. Geobey and Miss Julia Varley promptly proceeded to the disaffected centre, and arrangements were at once put in train for the organization of the strikers. The girls are typical factory workers, their ages ranging from fourteen to over thirty, and among them a general feeling of dissatisfaction appears to exist with regard to the rates of pay prevailing.

> Individual cases were brought to the notice of a "Gazette" representative, which were stated to be fairly typical. One woman was said to have to attend to three machines for a wage of 12s a week. A girl of sixteen said she could only average 4s 9d to 5s a week, and another, twenty-seven years of age, said her day rate was 7s a week, and at piecework she could not always earn that amount.

> This seemed to be a general grievance, that the girls could not obtain at piecework as much as their day work rating, and they are almost always wholly engaged on piecework. The remuneration varies with the character of the employment, the hinges having to pass through seven or eight different processes. One girl said she had only averaged 5s a week, though rated at 7s, and another girl of thirteen only got from 4s to 5s a week. The case of a girl, who has to pay 6s a week for board and lodging, and had not been able to earn that amount for months, was mentioned, and also that of two sisters, one of whom frequently had to give her sister a shilling or two at the week end to make up her wages.

> Another of the complaints of some of the girls is that at the end of the week they are not paid for what they have actually done, but for the amount of their work

that has been packed for dispatch. It was pointed out that this involved considerable hardship in some cases, as it frequently happened that girls had to wait for weeks and even months before some of their work was packed and they got paid for it.

Instances were given of girls who had left the works and not obtained some of the money they had earned. The strikers are demanding as one of their conditions that all work shall be paid for when finished.

Although the female hands at Messrs. Guest, Keen, & Nettlefold left work in sympathy with the strikers at United Hinges, they are also bent upon obtaining an improvement in the conditions of their employment. It is stated that their work is of the most laborious character, but the rate of payment is not commensurate. Girls engaged in painting bolts said they only averaged about 6s a week, although in engineering works men are employed to do exactly the same kind of work at much better wages . . . about a hundred girls are employed at the Storr Valley Works, all of whom came out on strike, and they were accompanied by sixteen men and boys engaged as tool fitters. They made the taps and dies for the machines on which the girls work, and they came out in sympathy with them.[90]

While the men's grievances boil down to "a minimum wage for all adult male workers the district rate of wages for all skilled operatives," the women's demands were more extensive, involving: a 10s minimum; a guarantee that "all work shall be paid for when finished," and, it would also appear, equality with men who were "employed to do exactly the same kind of work at much better wages."[91] We take the absence of additional demands by the men here, or later, compared with the more extensive demands of the women, as indicators of the sex-caste system in place at these firms. We have no evidence of male workers complaining about their wages being withheld, and we would suggest that the Kenricks probably would not have tried getting away with such a transparently exploitive practice with the men.

In the statement attributed to the women workers at Guest, Keen, & Nettlefold, we find an explicit reference to, and dissatisfaction with, sex-based wage discrimination.[92] The statement is tantalizing. It suggests, at the very least, that these women workers were well aware of, and were dissatisfied with, their subordinate position within the industrial sex-caste system. It also makes us wonder to what extent such resentment might have provided a good deal of the fuel for the "smouldering fires of discontent" to which the *Gazette* had referred. But what we do not have to wonder is whether the WU had helped create or would tap into this potentially rich reservoir of women's resentment, because the evidence shows that this grievance is ignored as the WU goes about institutionalizing, rather than challenging, the sex-caste system. In his history of the WU, Hyman writes,

The strike quickly spread to the firm's female workers and to neighboring factories. The Workers' Union had recently started organizing at these factories, creat-

ing the enthusiasm which led to this spontaneous outbreak . . . Beard and Varley
took charge of the dispute and found that the women's grievances were the most
serious.[93]

As we might expect, Hyman gives the WU and its organizers a prominent place in
the story, and so all the action in his account, aside from the "spontaneous outbreak," is initiated by the WU and its organizers. As suggested above, challenging
sex-based wage discrimination had something to do with the "spontaneous outbreak" of the women. But at the same time, "taking charge" of the dispute also
seems to have meant smothering that aspect of the women's "spontaneity."

What is even more interesting is how Beard and Varley "found that the
women's grievances were the most serious." Here, the women are the focus of
attention, but in a very specific way. What is "most serious" about the women's
grievances is that they are poorly paid, but as the organizers will go through great
pains to avoid saying throughout the strike, *not in relation to the men.* The idea of
Beard and Varley "finding" the women's grievances is an interesting gloss as
well. Women were the initial focus of the *Gazette* story about the strike, in part,
because the idea of women working outside the home, particularly in factories,
disturbed the middle class Victorian sensibilities of the *Gazette's* readers, and was
a hot topic throughout the period.[94] But women were also a focus of the story
because Julia Varley *made* the women's interests the focus of attention. But, of
course, as an organizer for the WU, she constructed these interests in particular
ways, and with particularly important consequences.

Julia Varley came to Birmingham in 1909 to become secretary of the Birmingham Women Workers' Organization Committee at the invitation of the social reformer and businessman, Edward Cadbury. She also was a member of the
National Federation of Women Workers (NFWW), and formed a branch among
the cardboard box makers at Bournville. After successful campaigns for the NFWW
in the Birmingham baking industry, and with women chain-makers in Cradley
Heath in 1910, she was asked to collaborate with the WU during the Bilston metal
workers' strike of 1911, and subsequently joined the Workers' Union. Varley was
asked to organize for the WU because of the high concentration of women workers in the engineering and metal trades firms in the area. Believing that when
women and men worked together they should be organized in one union, Varley
left the NFWW and became the first woman member of the WU's staff in 1912.[95]

Varley's strategy of agitating among and focusing attention on the women
workers apparently began well before the strike. Soon after her arrival in Birmingham in 1909 she was appointed to the Executive Committee of the Birmingham
Trades Council, and no doubt was the driving force behind the Council's investigation of women's wages and working conditions (noted in the *Gazette* story previously).[96] Moreover, it appears that the committee and Varley had specifically
targeted the Kenrick firm. In their account of the WU activities in the region during this period, Leask and Bellars write:

A Union enquiry at that time revealed that at Messrs. A. Kenrick and Sons Ltd., 120 women earned between 6*s* and 8*s* a week; 50 earned between 8*s* and 10*s* a week; 64 earned between 10*s* 6*d* and 12*s* a week; 29 earned between 12*s* 6*d* and 14*s* a week; and fourteen women of outstanding capabilities earned between 14*s* 6*d* and 21*s*.[97]

Thus, given what we know about Varley, the WU's organizing strategy, and the "investigations" of Kenricks already conducted by union organizers, Hyman's story that Beard and Varley *"found* the women's grievances to be most serious" (our italics) tends to obscure the role that these organizers played in constructing these grievances, since it was largely through their agitating and investigating that these grievances, *as such*, were there to be discovered.[98] Clearly, these investigations also might have "discovered" that the women were interested in getting paid the same as unskilled male workers. That they did not, points to the particular way in which the WU *constructed* these women's interests.[99]

Varley's belief that women and men who worked alongside each other should belong to the same union sat rather uneasily with the WU's refusal to challenge sex-based wage discrimination. Mirroring how the Kenricks incorporated female workers into a sex-caste system at their workplace, the WU strategy included women within the union but then subordinated them once inside. Once the women's interests were constructed and shaped in ways that did not pose a threat to male supremacy, the men were re-enlisted to "protect" them, effectively reasserting male authority. Consider these excerpts from speeches at rallies given by the organizers during the strike:

Julia Varley spoke and said that the ability of the owners to get as much from the workers was "due to the fact that the working classes had been so careless to their own interests (applause). They had allowed the employers to get every possible penny out of them, and they had heard with pain from time to time that the people of West Bromwich had allowed their girls to work for 4*s* a week." She then went on to exhort the men: "and she urged that the men had enormous power in their hands to make the lot of the women better. They wanted them to enthusiastically take up the fight which was a fight for humanity. It was because they loved their own people and their own class that they were out on that mission, and they were glad the workers had risen in their might to alter the condition of things."

Then Mr. Geo. Geobey, organizer of the Workers' Union, addressed the crowd:

Look at the girls who were employed at Kenrick's, those who were working in the nickelling shop. They saw girls 14 and 15 years of age wearing huge clogs wringing wet from the time they went in to when they came out at nights. Those were the future mothers slowly but surely having their life's blood sucked out of them, and being made unfit for the burdens of motherhood. In the future these girls working under such conditions continually would probably have rheumatism or some other disease as a result of it, and if the men of that district asserted their manhood to its fullest they would say without any hesitation that they were going to be responsible for those girls' upbringing, and they were to wake up to their responsibilities. The fathers of the girls were assisting the employers to do what they were in permitting their girls to work for 4s, 5s, or 6s a week, out of which they usually received 6d. for pocket money. By allowing that fathers were subsidizing the employer's wages and he urged that they could not be engaged in a more glorious fight than to try and improve the conditions of those girls (applause).[100]

We have another instance of this practice of refitting patriarchy to the changing nature of capitalism in Jack Beard's Presidential address to the WU in 1916. In this speech, Beard devoted considerable attention to the question of "Women's Labor," saying, among other things, that:

> the community has no right to interfere with the occupations of women only on the grounds of motherhood (hear, hear). And whilst we may lay it down that in the interest of the race it is the first duty of women to care for the young, and that the earnings of the men must not be lessened by the entry of the women into industry, there should be no barrier of sex . . . The Workers' Union has already put aside obsolete methods of the past, and has set about the work with energy, and to a fair extent has induced the sexless spirit of organisation which has brought together men and women in the same Union (applause)."[101]

And, finally, in the celebratory address given by the WU organizers to the strikers in *The Birmingham Gazette,* 25 April 1913:

> officials of the Workers' Union were pleased with the solidity that the men and women had shown in the present strike, and it was hoped that the workers, although they had secured a victory, would not go about crowing over it. It had been a clean fight. The firm had not been attacked as individuals, nor had the strikers been dealt with individually by the firm. They would go back to work as an organized body of trade unionists, and it was the duty of the men to see that not only their own, but the women's scale of wages was maintained.

In each instance, we find the men "called upon" to defend the interests of the women workers, but only those interests as they were constrained and constructed upon patriarchal assumptions. The men were to enlist in the noble cause of making sure the women receive what they deserve, but nowhere is it suggested that the women deserved to be treated the same as men. Indeed, perhaps mindful of their role in stimulating this "unrest of women," the WU then was faced with the problem of reminding and reassuring everyone, but particularly the male workers, that the rebellion was directed at the employers and not, despite occasional appearances to the contrary, at *them*, that is, at men, or the patriarchy within which their sex provided them with privileges vis-à-vis women.

What we also find in the record is, over time, the gradual displacement of the women from the center to the margin of discourse about the strike. As noted above, in the very first account of the strike to appear in the newspapers, the women strikers were given center stage. But as the strike went on, the women and their concerns were marginalized. Indeed, by the end of the strike, the dispute had been entirely reframed as a battle between the Kenricks and their "men," with the resolution of the "girls'" issue tacked on as an afterthought. On 25 April 1913, *The Free Press* began its summary of the resolution of the strike with:

> The strike at the works of Messrs. A. Kenrick and Sons and the United Hinges was terminated last Saturday, when terms of the settlement were arrived at a conference between the management and the men's representatives. After the conference the firm issued the following statement—"The firm agrees that the rate for adult labor shall be raised to 23*s* per week of 53 hours; that the scale of day work wages for girls shall be the same as that granted at Guest, Keen, & Nettlefold Ltd, Stour Valley Works."

Aside from Askwith, secondary accounts of the strike have tended to place the women on the same trajectory as in the newspaper accounts.[102] Thus, Leask and Bellars write:

> During the time of the Black Country strike, Julia Varley was outstanding in her gift of meeting the strikers' wives and women associated with the strike, explaining to them the importance of the fight their men were making, and turning them into active supporters.[103]

Clearly, Varley and the WU negotiated with the women strikers what would come to be recognized as the "women's grievances," and so we are not suggesting that the women's interests were completely imposed on the rank and file by Varley, the WU, or history. But at the same time, we also have evidence that the women might have been open to a much more militant stance regarding wage inequalities than the WU was willing to admit or pursue.[104]

What seems likely, however, is that the WU was only committed to improving the conditions of women to the extent that doing so did not threaten male supremacy either in the workplace or at home. No doubt afraid to alienate the large number of unskilled men who they believed were their "natural" constituency, the WU backed away from gender equality. Yet, it is far from clear that such a position would have alienated *all* unskilled male workers. As the *Gazette* account of the Guest, Keen, & Nettlefold workers makes clear, the sixteen men and boys who tended the machines upon which "the girls" worked "came out in sympathy with them." And these were "girls" who were demanding to be paid equal wages when they were doing the same work as men.[105]

Finally, it is important to recognize that by entering into an agreement with the unions, the Kenricks and other Midlands employers effectively put an end to the militant, disruptive, and threatening tactics of these unskilled workers. Indeed, we can easily imagine the Kenricks eventually recognizing the wisdom in the WU pitch that employers who allowed the WU into their works were less likely to face the disruptions of strikes.[106] During the strike-plagued years of 1911-1913, avoiding such disruption had become a primary concern of employers, particularly in the engineering industry.[107] The agreement, dated 7 July 1913, between the WU and the Midland Employers' Federation (MEF) alluded to in Sheila Lewenhak's account, recognized the WU "as an appropriate body to negotiate on behalf of the laborers and semiskilled workers, and a negotiating procedure was established." That negotiating procedure is listed in a section entitled "Provisions for Avoiding Disputes," under a subheading entitled "Breeches of Agreement": "This agreement is entered into on the understanding . . . that the rules of the various Unions involved efficiently deal with breaches of agreement by their members and that the rules in such cases will be enforced."[108]

Thus, from this point forward we find the employers in a position to hold the unions responsible for disciplining their workers. Commenting on this agreement after World War I, Askwith, too, notes how important the agreement was to dampening industrial unrest:

> This agreement . . . was remarkable for the success of its provisions for avoiding disputes, whether on general rates, piecework, and sectional rates, or sympathetic strikes. It created quite a new spirit in the Midlands, and months afterwards both employers and workmen informed me how successful it had been. It created order out of a very chaotic condition; the strike perhaps proving a blessing in disguise, because it provided methods for dealing with difficulties which proved of service during the War.[109]

But the question we might pose is this: to which "difficulties" was Askwith referring? Was he concerned primarily with difficulties caused by striking workers, or difficulties caused by *women*—who by 1918 constituted a large percentage of workers in the metal trades (60 percent of the Kenrick workforce, for example)? We would suggest that, among male factory owners and government officials, the desire to control industrial unrest might have been given extra impetus in the case of women workers who, with their militancy, threatened both profits and patriarchy.[110]

A "Victory for the Workpeople?"

Conventional accounts of the Black Country Strike have paid little attention to gender or to patriarchy.[111] From our vantage point, it is clear why the Black Country Strike came to be seen as a "victory for the workpeople" and a defeat for the employers. With patriarchy, or male centrality and authority assumed, the contest between the Kenricks and other Midlands metal trades employers and the Workers' Union was presented as a stand-off between "employers" and "workpeople," with the "workpeople" victorious. But, including the women workers in the picture and examining the outcome of the strike from their perspective, as we have done here, suggests a more complex result.

When all is said and done, neither the strike nor its resolution represented a significant challenge to male authority and privilege in the factory or at home.[112] Yes, the Kenricks and other Midlands employers were forced to bargain with the WU, and the agreed-upon scale of minimum wages certainly lowered profits and limited their ability to exploit unskilled labor, but their substantial investment in and dependence on the profits derived from sex-based wage discrimination were preserved. The rate of 23s established for the unskilled men, while by no means generous, represented an increase for many men and also established a floor below which they could expect not to fall. Moreover, the sex-based wage differential institutionalized the previous custom of paying women less than men, and thus preserved male privilege and superiority at the workplace and, no doubt, encouraged its continued practice in the home—at least in the short term[113] For these men, the strike affirmed their superior status as men, while it enhanced, ever so slightly, their status as workers. In this sense, it could be argued, the strike and its resolution deepened their commitment to capitalist patriarchy.[114]

As discussed above, we did find evidence that at least some female and male workers, particularly early in the strike, were ready to challenge the patriarchal foundation on which the Midland metal trades employers had built their capitalism and with which, despite rhetoric to the contrary, the WU was intent on building their trade unionism.[115] The call, not only for the establishment of a minimum wage, but for *equal* minimum wages for male and female workers engaged in similar work, had it been advanced by the WU, would have challenged a significant source of employer profits and a cornerstone of working class male privilege. So, perhaps expecting stiffer or even insurmountable resistance from the employers, and weaker support or even hostile rejection by male workers, the WU did not take up the case for sex equality. Indeed, it appears that, even though the WU was interested in gaining the support of women workers and creating, as Jack Beard put it, a "sexless union," they did this by shaping the women's interests within a patriarchal framework. And so we find the WU not advancing the case for sex equality, but in fact ignoring or repressing such interest when it was expressed by workers.

While the male-dominated WU and the employers jockeyed for position within a discourse of "class," their mutual interest in control over the labor of women, in the factory and in the home, provided the basis for compromise. Thus, to the extent that neither the employers nor the WU challenged the sex-caste system in the factories, we see patriarchy helping these male-dominated organizations transcend the "class conflict" that otherwise divided them. But while we focus our attention specifically on the WU, this union was more typical than exceptional in its willingness to place women workers beneath men. At the 1913 Trades Union Congress (TUC), Ben Turner of the Woolen-Weavers rose to offer a resolution for a minimum wage for "all adult workers, and especially women . . . sufficient to provide the standard of life and comfort which they have a right to expect from their labor." And, as Lewenhak reports, "Congress cheered in response, and carried the resolution unanimously." Unfortunately, what at first might appear to be a strong statement for sex equality on the part of the TUC was transformed into a strong statement for sex *inequality* somewhat later when "a resolution for a minimum wage for adult *male* workers only," 50 percent higher than that asked for *all* workers, also received the blessing of the 1913 Congress.[116]

As is evident with both respect to the Kenricks and the WU, many decisions about "political economy" were informed by assumptions about differences between men and women. As the Kenricks and the WU constructed and pursued their interests in production, concerns about the unequal place of men and women in society never seemed far from their thoughts, lips, or deeds. Specifically, both the Kenricks and the WU appeared willing to entertain a transition to a more "modern" system of industrial relations (i.e., a future), as long as that system continued the subordinate role of women in England.

Assuming we are correct in our interpretation of these events, what we have described is one occasion when organized capital and organized labor found common ground on the oppression of women. This finding raises more general questions about the relationship between capitalism and patriarchy. Sylvia Walby has argued, in fact, that patriarchy and capitalism make for an unstable combination because capitalists and working class men have conflicting interests in controlling and exploiting the labor of women. Just as we observed with the Kenricks, capitalists want to exploit the patriarchal devaluing of women in order to pay lower wages and undermine the power of working class men. Working class men, on the other hand, who in patriarchal families benefit from women's work at home, want to preserve women's labor power for themselves. Walby sees these two interests as inherently conflictual, although she grants that capitalists and working class men have shown a willingness to resolve this conflict with periodic "political" compromises, that often involve the mediation of the state.[117] Perhaps the resolution of the Black Country Strike provides an instance of such a compromise. The WU did not, as many craft unions did, attempt to exclude women from the workplace.[118] Instead, they adopted a "progressive" policy of welcoming women into "modern industry." In this sense, one might say, they were willing to cede a certain degree of control over women's labor to employers, in exchange for which they were able to increase their own wages, maintain their economic superiority over women workers, and ideologically, at least, reassert their right to control women's labor in the home.

Shaping Struggles: Paternalism Undone, 1892-1913

At its peak, we have argued, paternalism was a potent force in neutralizing working class resistance. When, in the late 1880s and early 1890s, skilled workers got in the way of the Kenricks' plans to reorganize the shop floor, the owners simply replaced these skilled craftsmen with machines and with mostly "unskilled" women workers. And, as we have shown, they did so by drawing on their status in the local social hierarchy; by asserting a new morality of economic rationalism; and by exploiting the economic dependency of their workforce. As a result of this *real* subordination of craftsmen, the technical efficiency of the machines, and a high demand for their products, the years between 1893 and 1913 were the most profitable in the firm's history.[119] Distracted, perhaps, by their own prosperity, the Kenricks and other Midlands metal trades employers took little notice of deepening impoverishment and growing industrial unrest in the district, and they were caught off guard by the strikes of 1913.[120] In the months that followed, the initial wave of strikes that overtook Kenricks rolled through the Black Country, gathering strength and size, until by July hundreds of factories had been hit and the number of workers on strike or locked out on any given day had reached 40,000.

These strikes revealed a fundamental contradiction between the Kenricks' strategy of mechanization and the firm's scheme of labor discipline rooted as it was in paternalism. By treating the increasing number of "unskilled," female workers as inferior to male workers, the tradition of paternalism, with its reciprocating worker loyalty, was weakened to the point that the mass of workers were willing to turn away from their employers and toward the Workers' Union. Paternalism, as practiced, was simply ineffective at binding these workers to the workplace. In the Kenrick case, we would go even further to suggest that the essence of the paternalist tradition, revolving as it did around an unequal, but mutually respectful relationship between the Kenrick "masters" and the subcontractors, was *a bond exclusively between adult men*. In short, what we would suggest is that the essence of Kenrick paternalism was an adult male solidarity and supremacy that helped mitigate inevitable conflicts between employer and workmen. Unmodified, such an approach was bound to be less successful with both teenage boys and poorly paid unskilled men, and of not much use at all with young girls and women. As Lown points out, a distinctive feature of this brand of paternalism was its strong association with a domestic ideology that "stressed the primary identification of women with home and children and men as breadwinners, and its relationship to a legal and political system which marginalised and subordinated women's access to property and citizenship rights."[121] Clearly, by hiring large numbers of women—even "factory girls" whom, most assumed, would only work until they were indeed married—the Kenricks were, unwittingly, undermining the basis of their own political regime. Although previously paternalism had helped to suppress working class resistance, we find that, with the changing composition of the firm's labor force, the owners resorted to a form of patriarchal despotism that actually provoked working class antagonism and thus facilitated the breakdown and transformation of factory politics.

The employers had little experience dealing with organized, unskilled workers and were, at the outset, disorganized in their response. We saw this clearly in the Kenricks' case, where, alone, the firm was unable to repel the rapidly growing and militant Workers' Union when it struck the firm in early April 1913. But, apparently realizing the benefits of solidarity, by June, seventy-one employers had come together to form the MEF.[122] It seems that their "sole purpose . . . [was] to strengthen resistance to the union."[123] Repeating the Kenrick strategy, the MEF initially took a hard line with the Workers' Union, refusing to negotiate while the strikes continued.[124] But their strategy, ultimately, was no more successful; bowing to public pressure, stiff rank-and-file militancy, and the persuasive efforts of the government's chief industrial mediator, George Askwith, the MEF came to the bargaining table with the union and eventually agreed to the central demand for a schedule of minimum wages on 7 July 1913.[125]

So, while the employers initially met the Workers' Union with fierce hostility and resistance, they quickly changed their tune and learned to live with, and, to some extent even embrace, the unions. Understanding this shift—from resistance to the unions to controlling the unions—as well as the contradictions of these new "bureaucratic" factory politics—is the purpose of the next chapter.

Chapter 5

"A Personal Interest in the Prosperity of Their Employers": Bureaucratic Hegemony and the Origins of Social Patriarchy, 1914-1922

Labor troubles erupted once again at Kenricks within months of the settlement of the Black Country Strike, and, as in the spring, the conflict involved how the Kenricks were treating their women workers. Based upon an interview with workman Jack Jones, in 1967, Church writes:

> In October 1913, workpeople at United Hinges were again called out on strike, the dispute arising over a shop foreman whose dictatorial and bullying behaviour the women resented. Encouraged by Jones's success, in April they urged him to try to improve labour relations by drawing the attention of the Kenricks to the foreman's behaviour. After an unsympathetic hearing from John Archibald, a strike was called in protest against the foreman's action; Kenricks issued the men's leaders with dismissal notices, for the strike was unofficial and the union men thus vulnerable to proceedings for leaving work unfinished. One night shortly after this episode, against the advice of the more responsible members, two men "drunk with a purpose" left the pub where the union customarily met, broke into the works, severed the hose of the fire engine, and set fire to the premises. Both men were convicted of arson causing damage to the value of £1,854, and were sentenced to sixteen months hard labor at Staffordshire Assizes in February 1914.[1]

The story has, by now, a familiar ring to it: the workers come forward with a grievance, the chairman is "unsympathetic," and, in response, the workers rebel. The ending to this story is different, however, from that which occurred in April, and the difference hinges on the refusal of the Workers' Union to back the strikers—John Beard, District organizer of the WU, having "advised the men to go back to work."[2] Enraged no doubt at both the Kenricks and the union, the two men lashed out, causing damage to themselves, the Kenricks, and, if we can believe Jones, to the union as well.[3]

Yet, fast-forward seven years and we find the Kenricks dealing in a rather different way with their workers. On 24 March 1920, Clive Kenrick, Works Manager in charge of labor relations at the firm, joined the four other members of the Management Board of the Birmingham District Engineering Trade Employers' Association at the Phoenix Hotel in Birmingham for their bimonthly meeting. Included with other mundane matters discussed, the secretary recorded that:

> Mr. Clive Kenrick reported that his men employed in the foundry had submitted a request to commence work during the summer at 6:00 am and finish at 3:00 pm with a break about 11:00 am for a meal, and it was decided that the application should be refused and that the men should be asked to refer their request through their union, who should approach the Association for a Conference.[4]

How did it come to pass that by 1920 we find a senior director of the Kenrick firm seeking guidance and recommendation from the Management Board of a Birmingham employers' association about a relatively trivial matter at his factory? Moreover, how did it come to pass that, instead of giving his men an immediate and direct answer, as John Arthur Kenrick had done seven years earlier, Clive is instructed by the Board to, in turn, instruct his men to take their request to their union representatives, who in turn, are supposed to request a conference with the employers? And how did it come to pass that everyone involved viewed such a "cooperative" and bureaucratic approach to labor relations as utterly normal?

Comparing Clive's response to the "summer hours" request, in 1920, to John Arthur's behavior in the "arson" incident, seven years earlier, suggests that shortly after the Black Country Strike, in the early days of "industrial relations" in the Midlands, the directors' approach to dealing with workers remained crudely autocratic. Their bureaucratic skills were as yet undeveloped, and the practice of feeding worker requests, concerns, and grievances into the "machinery" of negotiation was not yet acquired. To put it another way, John Arthur was an autocrat while Clive was a bureaucrat, and what we will try to do in this final chapter is attempt to understand how and why the firm's approach to labor changed from autocratic to bureaucratic in this relatively short period of time.

Our focus, of course, is not personality, but political economy. Encouraged by the state to cooperate with unions in peacetime, and forced by the state to do so in wartime, the Kenricks and their colleagues in the Midland Employers' Federation (MEF) quickly learned that there were advantages to working with, rather than against, the unions. Because, by this time, labor and capital were organized beyond the firm level, and because during wartime the state required industrial relations be coordinated at the national level, by the "summer hours" incident in 1920 the firm's labor policies were influenced at least as much by agreements brokered in Birmingham, York, and London as they were by decisions in the Kenrick boardroom in West Bromwich. And while the firm's directors would make every effort to steer those decisions made in far-off cities in ways that would maximize their managerial autonomy and further their economic success, what they did, or did not do, vis-à-vis their workforce was thereafter entangled in the politics between and within the key institutions of the modern, corporate economy: organized capital, organized labor, and the state.[5] In this chapter, then, we consider the rise of a third production regime at Kenricks that we call "Bureaucratic Hegemony."

The Kenricks and Organized Capital

When, in the latter part of the nineteenth century, employers in the Black Country metal trades deskilled their workplaces, hired unskilled male and female workers, and subjected them to routinized, low paying work, they created the conditions that would lead to a collective, class-based attack on those employers in 1913. And that attack, in turn, created the impetus for the employers to organize in defense of their interests. Nevertheless, many metal trades employers in the district, even some hit by strikes, chose not to join the seventy-one founding members of the MEF, suggesting that, in this case at least, it was no less difficult for employers to define and act on their collective class interests than it had been for workers in the district.[6] Indeed, the Kenricks themselves did not rush to join the MEF; as late as 13 August 1913, "on the question of joining a Midland Employers' Federation," the directors were willing only to "continue negotiations on the basis of the establishment of a satisfactory guarantee fund and management committee."[7] Eventually, however, the directors' concerns were addressed, and the firm signed on with the new organization.[8]

The MEF consisted of metal trades employers, such as the Kenricks, who employed large numbers of unskilled workers and women, and therefore felt threatened by the WU, but which were not, strictly speaking, engineering firms. A significant number of engineering firms in the district were already organized into a branch of the national Engineering Employers' Federation (EEF), which was known locally at this time as the Birmingham District Engineering Trades Employers' Association.[9] This organization's primary purpose was to represent the employer's interests against the Amalgamated Society of Engineers (ASE), engineering's foremost craft union.[10] But because many engineering firms also employed unskilled and semiskilled workers, in addition to fully skilled engineers, the Birmingham EEF was drawn into the Black Country Strike, initially instructing several of its members to raise the wages of the unskilled workers, but to refuse to concede the principle of a minimum.[11] However, they were unable to sustain their resistance to the principle of a minimum after the MEF began talks with the WU over that very principle. As a result, the Birmingham EEF joined in the negotiations with the MEF, agreeing to the 23*s* minimum for unskilled adult men, and soon thereafter, to the same schedule for "women and girls" the WU had won from the MEF. As the minutes of the Executive Committee for 20 August 1913 showed:

> Regarding the Workers' Union application for adoption of a schedule of rates for women and girls . . . This matter was again discussed when it was felt that in view of the action of the Midland Employers' Federation in agreeing to a schedule list of rates with the Workers' Union, this association was more or less compelled to make some kind of concession.[12]

It was a desire to avoid such concessions, and other labor problems that multiplied with the onset of war the following year, that led the two organizations to coordinate their labor policies during the war, and finally to merge in late 1918.[13] However, as we will discuss in more detail below, the fusion of the two organizations—one rooted in a struggle with a craft union and the other rooted in a struggle with the Workers' Union—would be a continuing source of tension within the expanded organization, leading in the case of the engineer's lockout of 1922 to open conflict and near schism.

Upon joining the MEF, several Kenrick directors soon acquired leadership positions within the organization. Clive Kenrick, Works Manager and younger cousin of the firm's chairman, Sir George, was one of four MEF representatives to meet with the Birmingham EEF in 1915 to coordinate the activities of the two groups during the war, to serve on the MEF committee that hammered out the amalgamation agreement in 1918, and to be elected to the five-member Management Board at the first meeting of the newly expanded organization on 17 February 1919, a position he would hold through the end of our story.[14] In addition, over the years, Clive would serve on numerous standing and ad hoc committees, and he would be the point man for the organization on the question of "female labor." From these positions, he devoted countless hours and energy to aligning the interests of the firm with the interests of the organization, and pursued those interests in struggles with both organized labor and the state.

The Kenricks extensive involvement in the MEF and, subsequently, the Birmingham EEF, suggests, on their part, an emerging class consciousness.[15] Forced into dealing with organized labor, the firm joined other similarly situated employers to face the unions. But while it is crucial to recognize that after 1913 the Kenricks developed a certain consciousness of class, it is also important to keep in mind that they continued to identify themselves as an independent family firm of hollowware makers. Understanding that the Kenricks had such multiple, and sometimes conflicting, identities and loyalties, alerts us to the possibility of political struggles within the EEF, and allows us to trace the impact of these struggles on labor relations at the level of the factory.[16]

State Intervention

The "arson" incident at Kenricks in the fall of 1913 illustrated to someone like Lord Askwith, the government's Chief Industrial Commissioner and mediator of the Black Country Strikes, everything that was wrong with industrial relations in Britain in the early years of the twentieth century. At the root of the problem was a mistreated and discontented working class looking for a measure of respect from their employers and society. Next came employers who responded to worker grievances unsympathetically, and eventually, punitively, making matters worse, rather than better.[17] For their part, the unions were often too weak or too unorganized and thus unable or unwilling to effectively represent workers. And left feeling powerless, alienated, and angry, there was concern that workers might, as the two at Kenricks had done, turn down a path of self, and, ultimately, societal destruction. For those who either assumed, or were assigned, the job of addressing these problems, the trick was to fix what was wrong with industrial relations while preserving British capitalism.

Given the British preference for a "night watchman" state, reform came not as a master plan, but in piecemeal fashion over a period of years.[18] Nevertheless, the thrust of the state's intervention in the economy around the turn of the century was to provide the working class, particularly those at the very bottom, with a minimum level of social and economic security, and to discourage strikes and social unrest by encouraging the peaceful settlement of industrial conflict through "conciliation." In short, the goal of state intervention was to level the playing field somewhat, and provide new rules for the game played on it.

Predictably, employers insisted that whatever the government was proposing to do in the way of reform was either unfair, unnecessary, unworkable, or perverse, and usually all of these things. For example, as the Black Country Strike occurred, the Board of Trade was in the process of extending the Trade Boards Act of 1909 to the hollow-ware industry, which would have established minimum wages. Undeterred by the strike or its settlement, the Board acted within a month to "schedule" the industry. News of the impending requirements prompted a local newspaper to seek out the opinion of "a member of a large firm of hollow-ware manufacturers" for his opinion on the matter, to which he replied:

I for one cannot see the necessity or the desirability of scheduling the hollow-ware trade as a sweated industry. So far as I know, there is no employer in the trade who is now paying less than the scale which has been agreed upon with the men's associations. The advances in wages recently agreed amount to a considerable percentage, and I have not heard of a single firm refusing to pay the prices.

If there is any firm paying less, then the work-men's associations have not carried out the undertaking they gave when the advances were made, for they then
agreed to withdraw the employees from any works where the increased prices
were not paid. Our firm, and I believe also the other firms in the district, are
paying higher wages than are being paid by manufacturers' organizations for
various other trades in the Birmingham and Black Country area, and if the employees are engaged in those trades are satisfied with the wages paid, those employees in the hollow-ware industry ought to be more than satisfied with what we
are paying. You may possibly be aware that the piece work wages in the galvanized hollow-ware trade are paid to a schedule which the men have drawn up.
Their demands have been granted in full, though in many cases they were unrea-
sonable. Whatever justification there might have been for scheduling the trade
before the new arrangement was entered into between the employers and the
men, it has now entirely disappeared. The position has been reversed, and while
at one time workpeople did not receive sufficient remuneration, now employers
are paying a great deal more than they should be called upon to pay, with the
result that sales have fallen off most materially . . . I don't see how the scheduling
is going to help anybody. If it were enforced—and I don't anticipate it will be—
they could not, surely, call upon us in this industry to pay a great deal more than
many other industries are paying.[19]

He then went on to say that a permanent schedule of wages would be impossible
because of all the different articles that would have to be weighed and measured,
prices for each decided upon, and so forth, and that as a result, no employers
would be willing to spend their time on it.[20]

State-sponsored social insurance schemes were another way the state attempted
to establish a floor below which the poor could not fall. The introduction of a non-
contributory old-age pension scheme in 1908 (of up to 5s per week) was the first
of the welfare provisions, and enrolled over half the people in the country over
seventy. And by the outbreak of war in 1914, nearly one million pensioners were
receiving just over 5s, costing the state some £14 million per year. More significant, however, particularly for our purposes, was the national Insurance Act of
1911, because it was targeted specifically at the "working poor" who now made
up a significant proportion of the workforce at Kenricks and in the metal trades.
Part I of the Act provided health insurance to those making less than £160 per year
(£250 after 1919), quickly attracting millions of subscribers. Part II addressed
another persistent problem of working class life: unemployment. Initially restricted
to males making up about 12 percent of the working population, it was expanded
in 1920, and again in 1936, so that, by 1939, 65 percent of the working population
and their dependents were covered.[21]

While these various forms of social insurance did not significantly reduce the
dependency of working class people on wage labor, it did loosen this dependency
somewhat. With a small pension to look forward to, older workers might not have
had to work so hard, as had been necessary in the past; nor was it necessary for

them to work when ill to avoid a total loss of income; and when laid off work, they would receive something to tide them over.[22] Moreover, since these social insurance schemes were paid for, in part, through a payroll tax on employers, they resulted in a redistribution of income from employers to employees via the state— a "raise" if you will. And apparently this "raise" had a large enough effect on profits that it was worth noting, along with the impact of the increase in wages resulting from the Black Country Strike, in Kenrick's annual report for 1913, where we find:

> During the Spring, a strike without notice broke out in a small section of the works and fanned from outside gradually extended so far as to cause the complete stoppage of the works for a short time. Prolonged negotiations led to a settlement by which many workers employed at the lower rates of wages have received substantial advances. This and the full brunt of the Insurance Acts have added to the general costs.[23]

Taken as a whole, these state-provided insurance schemes made it marginally easier for workers to demand more, and put up with less, in exchange for their labor power. But at the very same time that these measures were in some sense empowering workers against their employers, they also worked to "buy off" dissent, channel whatever dissatisfaction that remained in a reformist rather than a radical direction, and legitimate the entire social system in the eyes of workers. And, as legitimacy rose, it was hoped workers might be more inclined to send their representatives to the bargaining table than to take themselves and their grievances to the streets.

While "Conciliation Boards" and "Trades Councils" were functioning in the more established industries—including, by the late nineteenth century, some sections of the engineering industry—to this point the state had done relatively little to encourage collective bargaining and rarely got involved in specific negotiations or disputes. But this changed with the Conciliation Act of 1896, wherein the state began to actively encourage the development of a system of industrial relations and to expand its role within that system. This act established a special section within the Board of Trade to:

1. inquire into the causes and circumstances of a dispute;
2. take steps towards bringing the parties together;
3. appoint a conciliator or board of conciliation on the application of the employers or the workers; and
4. appoint a conciliator on the application of both parties.

While still leaving the choice of negotiation up to the disputants, the Act encouraged organization and negotiation, rather than class warfare, and provided the procedures, personnel, and support for conciliation to succeed.[24]

By 1913, Askwith was the Chief Industrial Commissioner charged with conciliation work under the Act of 1896, and, as we have seen, had personally forged
the agreement between the employers and the trade unions that ended the Black
Country Strike.[25] Looking back on his role in the resolution of the strike, Askwith
noted how:

> It created order out of a very chaotic condition; the strike perhaps proving a
> blessing in disguise, because it provided methods of dealing with difficulties
> which proved of service during the war. . . . The scheme involved stages of ex-
> haustive discussion, speedy examination of claims, and no stoppage of work while
> negotiations were pending, together with avoidance of stoppage or suspension of
> work on account of outside disputes, while at the same time the right to strike or
> lockout was maintained.[26]

Although Askwith's "methods" would, indeed, prove of use during the war
and afterward, it would finally take the coercive powers of the state during wartime, and the appeal of lucrative munitions contracts in a declining hollow-ware
industry, to turn Kenrick autocrats into Kenrick bureaucrats.[27]

Controlling the Kenricks

The demand for Kenrick products dropped sharply with the beginning of war in
August of 1914, making it somewhat easier to let go, with a promise of employment upon their return, several hundred young men who within a year had volun-
teered for the trenches of France.[28] Within months, however, contracts with the
Ministry of Munitions for mortars, grenades, bombs, enameled water bottles, and
other items would more than make up for the lost demand for domestic goods.[29]

Excerpts from the firm's Annual Reports for the war years indicate the increasing extent to which the firm became dependent upon the Ministry of Munitions:

> 1915: Had it not been for large and profitable contracts from the War Office for
> enameled equipment the works at Stourport could not have shown so large a
> profit . . . The building trade—the backbone of Foundry work—has remained
> utterly stagnant and there is no prospect of recovery until after the war.[30]

> 1916: Since the last report your Directors have been able to further assist in mili-
> tary work through the opportunity given them by the Ministry of Munitions of
> supplying the large quantities required of both Mills Hand Grenades and Stokes
> Shells . . . Further Contracts have also been made for the machinery and equip-
> ment of plant, to be erected on the old works, capable of turning out ten thousand
> (10,000) enameled pieces per week, at an estimated cost of £10,000.[31]

1917: Owing to further contracts for Grenades, Bombs, Water Bottles and a number of smaller items required by the Military Authorities the actual total returns have been slightly increased, but the quantities supplied to our ordinary customers have been again reduced: while increased prices have tended to restrict demand.[32]

1918: The total output of the Companies has been fully maintained, but a larger proportion than before has been absorbed by the Naval and Military Services.[33]

For 1917 and 1918, military contracts accounted for 80 percent of the firm's business, with sales of £348,379 in the peak year of 1918.[34] Not surprisingly, the firm's "war boom profits" soared off the charts, and while the MEF grumbled about and won modifications to the "excess profits" tax levied on munitions makers, overall its members seemed satisfied with how they made out during the war.[35] As T. Harris Spencer, Chairman, stated in a report to the members on January 1917,

The Arbitrations before the Committee on Production have been 16, and of course, while we have had some varying successes there, I think we may claim the great majority of the Findings have been distinctly in favor of the Midland Employers' Federation.[36]

Although the firm made record profits, their workers took home sharply higher wages; however, the increase was offset to a great extent by the inflationary rise in retail prices, and it was only because of secure, steady full-time work, overtime, night-work, and bonuses that the real wages of Kenrick workers kept pace.[37] Nevertheless, the WU expressed satisfaction with the wage increases won for its members during the war.[38]

Crucially, however, the war profits and higher wages earned by employers and workers in the munitions industries did not come without strings attached. While government officials, such as Askwith, might be unhappy with the dysfunctional industrial relations exhibited in the "arson" incident at Kenricks in the fall of 1913, politicians were unwilling to force cooperation between the employers and the unions during peacetime. But a year later, with a war on and the nation's survival at stake, the sort of feuding between employers and workers that had been going on prior to the war had to stop—voluntarily if possible, but by force if necessary.

Kenricks began receiving contracts from the Ministry of Munitions (MUN) in 1915 when it became clear to Britain's leaders that the government arsenals and the War Ministry's official list of munitions makers were unable to keep up with the supplies required of industrial warfare. To meet the demand, the MUN brought thousands of firms, including most in the metal trades, under its control. As a result, Archibald Kenrick & Sons Ltd. became Controlled Establishment number 1535 on 8 August 1915.[39] The rules that both the Kenricks and their workers were subjected to, and the changes they were required to make under the Ministry of War Act of 1915, were designed to prevent exactly the sort of outcome we observed in the "arson" incident prior to the war, and to encourage the sort of outcome we observed in the "summer hours" incident after the war.

The first thing to go was the strike. Because state managers realized that the cooperation of labor was essential to maintaining the flow of munitions, state officials vowed to treat the unions as an equal partner, and to consult with labor leaders about proposed changes in wages and workplace organization—particularly on the question of "dilution" (i.e., the subdividing and mechanizing portions of the labor process and the substitution of unskilled workers, particularly women, in place of skilled craftsmen).[40] In exchange, at the Treasury Conference in the spring of 1915, the unions agreed to abandon the strike for the duration. The Munitions of War Act, passed during the summer of 1915, superseded the union's pledge, and while not completely forbidding strikes and lockouts, it made it almost impossible for a legal one to occur. A union was only allowed to strike, or an employer only allowed to lock out, after the dispute had been reported to the authorities for referral to arbitration within twenty-one days. But since any award issued under the Ministry of War Act was enforceable by law, all the government had to do to prevent a legal strike or lock out was refer the dispute to arbitration.[41]

As a result, within controlled establishments, employers lost unilateral control over their workers and their workplaces. While effectively prevented from striking, workers now could take their grievances around "unsympathetic" employers and straight to Askwith's office for mediation or arbitration—an option that was previously either unavailable or irrelevant so long as the employer could ignore the arbitrator's findings. But under the Act, employers could not ignore the arbitrator's findings, and so the unions increasingly began to circumvent the employers. In fact, according to the Birmingham Engineering employers, unions were encouraged to do just that when the Ministry intervened in 1916 to raise wages without a request from the unions, pre-empting an existing agreement. Infuriated by the Ministry's failure to consult them, the Birmingham employers wrote to the Emergency Committee of the Engineering Employers' Federation, stating:

> The issue of Statutory orders Nos. 447 and 456 by the Minister of Munitions has placed the members of the Birmingham Association in an intolerable position for the following reasons:
>
> (1) An Agreement dated 23rd November 1915 between the Workers' Union and the Birmingham Association was in existence at the time that the orders were made establishing rates of wages for the District for Girls and Women 14 to 21 years of age, and there was in the district no dissatisfaction on the part of the employees with the rates of wages prevailing under that Agreement. Although fully aware of the existence of this Agreement the minister ignored it, and substituted a different and higher rate of wages, without consultation with the Employers' representatives of the district.

(2) The action of the Minister is causing unrest amongst employees in that they now endeavor to obtain advances in wages by going direct to the Ministry, or by bringing political pressure to bear on the Ministry, instead of by going through the usual channel, viz., applying through their Trades Union Representatives to the Employers' Association, thus upsetting unnecessarily the Machinery, which has been created with so much trouble and perseverance for friendly negotiation between Employer and Employees, and placing the Employer in the position of not having full control of his own works, although providing the wherewithal to run it.

(3) Another effect of the order which has already been felt to a serious degree is that the rates fixed for girls are in many cases higher than those fixed by agreement with the Workers' Union for boys and youths, which were at the time the orders were issued being amicably worked to. A comparison of their own rates by Boys and Youths with those fixed by the Ministry for girls has already resulted in trouble in that direction.

(4) The action of the Minister is also a violation of a principle which has been held by the Birmingham Association ever since it was formed, that no advance in wages should be given without an application for the same from the employees.

(5) The cost of production of various articles which are being made in the district at contract prices will be correspondingly increased and the difference in cost will be borne by the Employers who having no notice that such an order was likely to be made, have not provided for the contingency, and therefore will suffer serious loss.[42]

While the employers were obviously praising the existing negotiating "machinery" in this context primarily to highlight their objections to the Ministry's action, their letter suggests, nevertheless, that the Birmingham EEF was, by 1916, already invested in, and convinced of the need for, cooperation with labor. But while the Birmingham EEF appeared only willing to concede the need for cooperation, the Kenricks and their colleagues in the MEF were by this time enthusiastic proponents of cooperation, as evidenced in T. Harris Spencer's speech to the assembled members on 4 January 1917:

I find it is the general opinion that the best results will be obtained, and the best interests of the country will be served, by some cooperation between capital and labor. (Hear, hear) . . . What we have to get into our minds is this, that the conditions of the people of this country after the war will have to be much improved compared with what they were in pre-war times. I think, gentlemen, I could argue with you that will be one of the best propositions I could make in the interests of the employers . . . How is it all to be achieved? Not by misunderstanding, not by strikes, not by legislation, not by suspicion or ill will. None of these things will be any good if we are to achieve these great objects of the future. It can only be done as far as I can see, and as far as I want you to see, by cooperation.[43]

Given the fact that the MEF was founded less than five years earlier *explicitly* for the purpose of defeating the Workers' Union, this was a dramatic shift in attitude on the part of these employers. With its assertion that the employers' interests were best served by well-paid workers who were treated as equals, Spencer's speech could have been written by Askwith or the Minister of Munitions, or someone similarly charged with the overall well-being of the capitalism in Britain. Clearly, by this time, the employers of the MEF had absorbed and were reflecting back a good deal of the wartime corporatist gospel coming from the MUN (see below).

The shift from an ethic of conflict to one of cooperation and compromise appears less dramatic, or is at least easier to understand, when we learn that between 7 August 1913—the day the agreement ending the Black Country Strike was signed—and 4 January 1917—the day the above speech was given—representatives of the MEF had met successfully in conferences with labor leaders over four hundred times.[44] However much MEF committee members might have regretted the time spent on these meetings, four hundred meetings were certainly preferable to four hundred strikes or lockouts. As the months went by, and the number of successful meetings piled up, the employers must have realized that the leaders of the unions like the WU, despite their sometimes alarming rhetoric about "control of industry by the workers,"[45] and their street-level militancy, were not revolutionaries out to overthrow British capitalism, but only reformers intent on securing a bigger share of the economic pie for workers—a goal even the employers finally came to agree was worthy.[46] Experience, then, taught the employers what the WU leaders had claimed from the beginning: that those who worked *with* rather than *against* the union were less likely to be victimized by strikes and labor problems.[47] As prominent members of the MEF, with a record of committee service, the Kenricks were likely among the first of the employers to realize this.

We should not, of course, take the employers' expressed interest in cooperation completely at face value. It seems clear that after 1913 they grew serious about meeting with and negotiating with labor leaders, rather than trying to destroy the unions. Beyond that, however, it is also clear that the employers were willing to take advantage of the system. For example, if the unions controlled the workers, and the employers could control the unions, then the unions might be incorporated within an expanded factory regime of labor discipline. We see elements of this strategy in the "summer hours" incident at Kenricks in 1920. By bringing the question of changing hours to the EEF Management Board, Clive is not only acting as the loyal "organization man," not wanting to act unilaterally in a way that would undermine the collective interests of the Birmingham employers, but is also *using* the organization, the union, and the bargaining "machinery" available to manage his workplace. What is more, because the unions are included, decisions that come down from the "machinery" have a legitimacy they lacked when coming unilaterally from the employer—as in the case of the "arson" incident. This made it more difficult for unhappy workers to mobilize resistance to the decision, since such resistance had to confront both the employers and the union leadership.

Welfare Supervision and Wartime "Dilution"

Other wartime state interventions that were pushed on controlled establishments were various "welfare" requirements, which, interestingly enough, were funded from an excess profits tax placed on the employers. At Kenricks, these welfare provisions became compulsory in 1916; management agreed to set up first-aid stations at various places throughout their works as well as to appoint at least one woman as "welfare supervisor." In that same year, the Munitions minister authorized charging costs against profits for canteens at West Bromwich and Stourport. And, in 1917, the company would make an allowance of 10s per week towards the expenses of a "men's recreation club."[48]

These tactics were part of a Ministry of Munitions welfarist management ideology embodied in the formation of the Health of Munitions Workers Committee in 1915 and the appointment of Seebohm Rowntree as Director of the Ministry's new Welfare Department.[49] The committee was charged with studying the "personal health and physical efficiency" of munitions workers generally, yet focused much of its attention on women and children. Not only, it seems, was the state concerned with protecting the role of women in building healthy families and strong communities, but—as the Kenricks themselves had found some years earlier—also with the specific problem of disciplining the thousands of women workers in munitions plants. The issue of controlling women workers had been brought before the MUN during sporadic strikes by female munitions workers, the most dramatic of which was when nearly 6,000 women walked off the job at the Amstrong-Whitworth plant at New Castle-on-Tyne in March of 1916. By the end of that year, several hundred, mostly middle-class, "lady" supervisors had been installed in national munitions plants across Britain, reaching nearly a thousand by the end of the war.[50] "Controlled Establishments," such as Kenricks, were expected to hire welfare supervisors on their own accord, and the firm's records indicate such personnel were present during the war. Unfortunately, other than knowing welfare supervisors had been hired at the works, there is no surviving evidence regarding how the women were received by either the workers or the managerial staff, nor how effective they were at their assigned tasks. Yet, we do know from other studies and MUN documents how welfare supervision was discharged at other factories and can only assume that their role at Kenricks was not radically different.

Seebohm Rowntree, having himself employed women and girls for years at his family's cocoa factory, had, according to Laura Downs, a "corporatist vision of a reformed factory community, socially and spiritually regenerated by the healing balm of welfare management."[51] Rowntree believed that the role of the female supervisor was central to building that community and to eliciting the cooperation and consent of the female workforce. According to Downs:

> The ministry recommended that these women oversee the full range of welfare installations, such services as the canteens and infirmaries, whose benefits extended to male as well as female workers. But the heart of their work lay with the female workforce: hiring and deploying new workers, supervising the night shift, handling all breeches of shop discipline, keeping records of absentees and poor timekeepers, looking into complaints from workers, and investigating "slow and inefficient work or incapacity arising from conditions of health, fatigue or physical strain."[52]

For Rowntree, "If the welfare workers have the confidence of the employees, and are always in touch with them, they will naturally be the medium whereby matters occasioning dissatisfaction or misunderstanding can be investigated and put right." But moreover, "an increase in efficiency is important not only to the employers but also to the workers; for there cannot be a progressive improvement of wages unless there is a progressive improvement in methods of production."[53]

As some combination of forewoman and shop steward for the women workers, the positioning of the welfare supervisor is interesting in light of the charges of "abusive and dictatorial" behavior toward women leveled by workers at the Kenrick firm two years earlier. Clearly, the MUN believed that, throughout the metal trades, a "female" touch would provide a buffer between male supervisors and women line workers. Echoing the division of labor and authority in the patriarchal family, Cecil Walton, manager of a national projectile factory, characterized the role of the welfare supervisor this way:

> When men and women are employed in the same operation foremen and forewomen are essential. This is no more dual control than that of a father and mother. It is combined, more intimate, and broader control. . . . The foreman's chief responsibility is production, and the quantity and quality of the work. The forewoman's just care is for the producers. She must keep her squad up to the required strength, and each individual of it happy and fit. This is a natural division of responsibility, which makes for sweeter working and greater efficiency.[54]

Yet, the strategy was not altogether effective. As Downs suggests, the major problem these women faced was "carving out a sphere of independent authority" in the midst of an "entrenched male technical staff."[55] Management often resented the intrusion of the state into their shops, and women workers sometimes found the welfare lady's middle class standards of respectability overbearing. Welfare supervision did not survive beyond the armistice, although the topic was apparently still up for discussion among Midlands metal trades employers as late as April of 1918, where at a meeting of the Executive Committee on 22 April it was noted that a Joint Conference was held between the National Employer's Federation (formerly the MEF) and the WU to discuss "duties and functions of welfare supervisors. Another meeting planned, Ministry of Munitions representative invited to attend." Soon after the war, however, the Kenricks did reduce the standard number of hours worked per week based on concerns about the inefficiencies resulting from overwork and fatigue.[56]

One of the most controversial and widely discussed aspects of wartime state intervention centered around the "dilution" of labor. Early on, the War Office hoped to satisfy its manpower requirements for the army and navy through volunteers. As a result, many skilled workers went off to war, leaving the factories with a severe shortage of skilled workmen as the war dragged on and the need for munitions increased. When it became clear that skilled craftsmen were better used in the factories than in the trenches, such men were prevented from signing up or, later, being drafted, and some were even recalled. The loss of craftsmen, combined with the need for a dramatic expansion of munitions production, meant that skilled craft work had to be, at least for the duration of the war, broken down and rationalized, that is, "diluted," such that unskilled workers, and particularly women workers, could do the work. The craft unions, of course, and particularly the ASE, were very alarmed and concerned about the long-term consequences of dilution, and in fact only agreed to the no-strike pledge after receiving an assurance from the government that "pre-war practices" would be reinstituted after the war.[57]

"Dilution" was, of course, nothing new to Kenricks; they had been "diluting" work at their factory since at least the 1890s. In their required "Returns from Controlled Establishments" to the MUN, the Kenricks reported that they employed 306 (33 percent) female workers in July 1914, 493 (56 percent) in July 1918, and 549 (60 percent) by October of that year. Seventy-five percent of women working at the factory in late 1918 were over the age of eighteen.[58] Thus, this dimension of state intervention likely had very little affect on the firm during the war, beyond finishing a process begun years earlier, which was evidenced by a further jump, some of it temporary, in the percentage of women workers at the firm.

The MUN did surveys of controlled establishments throughout the war, but in light of the missing and incomplete data we have on some of the forms sent in by the Kenricks, we should use the data gathered with some caution. With this in mind, we can get some idea of the range of work done in hollow-ware firms during the war, as well as some sense of how dilution was affecting these firms, from a table compiled by the MUN based on its October 1916 survey of controlled establishments. In answer to an open-ended question, the MUN produced a table entitled "Processes in which women are replacing men." Table 5.1 lists the frequency distribution for the firms classified in the "Hardware and Hollow-ware" industry.

Clearly, without knowing the total number of hollow-ware firms under government control, or the total number of processes done in those firms, it is difficult to make much of these data beyond the observation that some processes in some

Table 5.1. Processes in Which Women Are Replacing Men in the Hardware and Hollow-Ware Industry

Process	No. of Firms	Process	No. of Firms
Aeroplane engine parts	1	Machine operating	2
Boring	1	Manufacture petrol cans	1
Copper Bands	1	Molding	1
Coremaking	2	Packing	1
Dipping	1	Pickling	1
Drilling	1	Polishing	1
Electric welding	1	Press work	2
Electric galvanizing	1	Safety caps	1
Exploder tubes	1	Screwing	1
Galvanizing	1	Shell work	3
Grinding	1	Soldering	1
Gun carriage parts	1	Submarine mine parts	1
Hammering dents	1	Tinning	1
Laboring	1	Turning	1

Source: There were limits to how much engineering firms were willing, or able, to "dilute" their works. Thus, early in the war, at a General Meeting of the Birmingham EEF on 7 September 1915, we find in the minutes a report on a "special meeting with Minister of Munitions representative A. Wharton Metcalfe" at which the association reported it had "diluted as much as possible and now needs 2,000 skilled men," "Minutes, General Meeting, September 7th, 1915," BDEA, MRC/MSS.265/B/1/3.

firms were undergoing "dilution." Given that most skilled work at Kenricks had been "diluted" by this time, the vast majority of the women added to the payroll during the war would have simply replaced unskilled males who had gone to war or taken on what was left of the few skilled jobs remaining at the factory after they had been "diluted."[59]

One final point about wartime state intervention concerns government restrictions on worker mobility. As labor became increasingly scarce, workers began selling themselves to the highest bidder, leading to uncontrolled wage inflation, inefficiency, and a critical shortage of workers in the munitions factories. Employers in the EEF and the MEF took the lead in squelching this trend by agreeing not to poach each other's workers or to hire a worker who did not have a letter from a previous employer releasing her or him from employment.[60] Workers detested the system, of course, comparing it to indentured servitude. Nevertheless, the MUN adopted the system and made it law, resulting in a few widely publicized cases of workers being jailed for subverting it.

For the employers of controlled establishments, however, restrictions on the mobility of labor were very much appreciated, making it easier to acquire and retain needed labor without paying exorbitantly for it. The Kenricks indicated as much in their annual report for 1916, where it was noted that, "Under the circumstances (being a Controlled Establishment—auth.) it has been easier to retain workmen who might have obtained other jobs, and there has been no difficulty in keeping all hands fully employed on profitable work."[61]

As Askwith noted in his memoir, the seeds of cooperation between the MEF and the Workers' Union were planted in the agreement that ended the Black Country Strike. And in the months after the agreement, and consistent with its provisions on avoiding disputes, the employers and the unions began to meet regularly to work through their differences. After the war began, Askwith, and the others charged with keeping the industrial peace while the nation was at war, used the agreement in the Midlands as a template for developing a broader, and eventually a state-mandated, policy for the war years. The state's policy, in turn, accelerated and supported the turn toward negotiation that the employers and unions had taken, such that by the end of the war we find the employers and unions, quibbles and gripes aside, basically pleased with what they had gotten out of munitions work, praising the benefits of cooperation, and looking toward a postwar world of both international and industrial peace.[62] In his memorandum assessing the contribution that employers' organizations were likely to play in shaping postwar industrial relations, Askwith sized up the MEF as follows:

A somewhat unique type of employers association exists for the district formed by Birmingham and the Black Country; is called the Midland Employers' Federation, and it is a body formed of employers engaged in a number of miscellaneous industries carried on in Birmingham and the surrounding district—such as wagon and carriage building, bridge-building and construction work, hardware manufacture, tube making and metal stamping, nut and bolt making, and a number of trades of various kinds. It is a recently formed federation, its origin being due largely to the strikes which occurred in the district in 1912, and it has not been in existence long enough to enable an opinion to be expressed as to whether it is likely, in view of the varied interests of its members, to stand the stress of severe industrial disorganization. Up to the present the Federation has succeeded in working amicably with the large number of trade unions which cater for the miscellaneous grades of workpeople employed by its members, and collective agreements have been arrived at with practically all of them. These agreements, it should be pointed out, have all been in the nature of increases of wages and of levelling-up wages and working conditions, and perhaps some of the success of the Federation is due to that fact; the occasion for proposing to reduce wages has not arisen since the Federation has been formed.

The labor policy of the Federation has, however, been framed on broad lines, and has been sympathetic towards the lower paid sections of the workpeople, and there is at present very well-marked confidence between the Federation and the Workers' Union, which is the principal union representing the unskilled labor of the district.

One of the main objects of this Federation has been to secure a reasonable degree of uniformity in regard to wages and working conditions in the various sections of industry connected with the organization, and the Federation may therefore be regarded as likely to view with favor proposals for collective and uniform re-organization for peace purposes. The Federation has a membership of about 250, and its members employ approximately 100,000 workpeople with an annual wage bill of £6,000,000, so that it is a body of great influence in the Midland area.[63]

As prominent members of the MEF, the Kenricks were in the thick of these developments and undoubtedly led some of them. And as we will see in the concluding section of this chapter, with the merger of the MEF and the Birmingham EEF in late 1918, Clive Kenrick brings the firm into an even more prominent position among the Midlands metal trades employers during the postwar period.

Kenricks, the EEF, and the Engineering Lockout of 1922

Church provides us with a nice summary of Kenricks' increasingly dire situation in the early postwar period, writing:

> The end of the war brought a brief boom in hardware sales, accounted for by the interaction of depleted supplies with pent-up demand, particularly from overseas buyers; but within three years of the Versailles Treaty the Germans had retrieved much of the lost ground in markets for hardware . . . Fortunately for Kenricks, German goods were prevented from entering the Australian market, but in South America, where the company had established direct trading companies, German and American products were ousting those made in this country. The rise in the importance of the British Empire as markets for hardware, and especially for hollow-ware, continued, though even these markets collapsed with the onset of the depression in worldwide trade in 1921.[64]

As the firm's fortunes declined, the directors began looking for ways to maintain profit margins and hold on to their shrinking share of the market. One approach, floated in 1918 by William B. Kenrick, involved pooling the resources and markets of the thirteen members of the Cast Iron Hollow-Ware Manufacturers' Association (CHIMA) into a new, more efficient company that would effectively concentrate, and constitute, what was left of the British hollow-ware industry. Laying out the justification for his plan in the preamble to his proposal, William wrote:

> For some time past the Cast Iron Hollow-ware trade has been suffering from a lack of demand due to various causes, the principle of which is perhaps the long continued stagnation of the building trade, which shows no sign of coming to an end. But apart from this cause, there has been during the last 20 years a gradual loss of trade due to the competition of cheaper articles offered for sale for use in the kitchen. Up to now, although the total trade has diminished, makers have been able to maintain prices at a remunerative level, but it is evident that a num-

ber of causes are tending to increase the cost, and it is doubtful if consumers will be able to bear any corresponding increase in price. Makers are therefore faced with the problem of having to sell their goods at present prices, or to suffer further diminution in their sales, while their costs are bound to increase.

Can any method be devised to meet this problem?[65]

Since there is no evidence that the company was formed, we can only surmise that the answers given to William's question did not include taking up his proposal. Yet, that one of the directors would even consider merging the company with its competitors in such a venture testifies to the firm's difficulties at this time—difficulties that would get far worse after 1921, at which time the firm would have to resort to lay-offs, "short-time" (working only three days per week), and wage cuts for everyone including foremen.[66]

While William Kenrick was considering reorganizing the company, Clive Kenrick was doing his part by continuing his wartime practice of representing the firm, and the hollow-ware trade, within the MEF. And, as noted earlier, Clive was a member of the MEF committee who negotiated the merger with the Birmingham engineering employers in 1918, and in early 1919 he was appointed as one of five members of its Management Board. Clive's work on the Management Board of the Birmingham EEF, between the end of the war and 1922, was focused in two main areas. Most of his energy was devoted to representing the organization in negotiation with the WU, particularly on the question of wage rates for females—an issue of increasing importance to his own firm. The second, and more politically sensitive, job that he assumed was leading an independence movement within the EEF on behalf of the several hundred "Allied Trades" employers.[67]

In Clive's work with the WU on behalf of the Birmingham employers, we see a concerted effort to keep wages low and minimize labor conflict. For example, when depression hit in 1921, the employers were intent on bringing down wages as rapidly as possible from their wartime heights. Thus, from late 1920 on, Clive led a regular series of meetings with the WU to ratchet down wage rates.[68] Naturally, the WU opposed these reductions, often leading the two parties to refer the matter to binding arbitration;[69] however, increasing unemployment made the WU effort futile.[70]

It is interesting to note, however, that despite the employers' dominant position during this period, Clive and his colleagues were not at all heavy handed about wage reductions. As noted above, they were often willing to submit their proposals to an independent arbitrator.[71] Moreover, perhaps recognizing that every proposed wage reduction represented another potential source of conflict, the employers set out to rationalize the process even further by tying wage rates to a monthly cost-of-living index published by the Board of Trade.[72] Once the unions agreed to such a system, reductions were automatic, rather than negotiable, much like a "premium bonus" system was with regard to piece rates. The system also was more efficient, in the sense that it reduced the need for conferences, or at least the need for prolonged negotiations.

Another, more fundamental, employer strategy for controlling labor and reducing the potential for class conflict was to subdivide unskilled and semiskilled
labor as much as possible, creating a hierarchical ladder of skills and wages with
many small steps. The point of creating such hierarchies was to accentuate trivial
differences between workers and to increase competition, individualism, and loyalty to the firm, while reducing or impeding solidarity among workers. While in
many instances such hierarchies were simply imposed on workers, in a remaining
transcript of one of the conferences between the Birmingham employers and the
WU, chaired by Clive Kenrick, we find this hierarchy itself was the subject of
negotiation. The conference was requested by the WU for the explicit purpose of
eliminating several of the steps in the wage ladder proposed by the employers in
an earlier meeting, and involved extensive discussion between the employers and
the union leaders over the reasons for and against each of the wage levels in question. Excerpts from the exchange between the employers and the union leaders
provide a feeling for the issues involved and the tone of the negotiations:

> *Chairman* (Clive Kenrick): I understand, Mr. Geobey, you have come this evening
> to finish off this agreement on base rates for the engineering and foundry trades.
> There are one or two items you will want to discuss.
> *Mr. Geobey:* That is so, Mr. Chairman.
> *Chairman:* Will you tell us what they are?
> *Mr. Geobey:* Well the questions that we desire to discuss with you this afternoon,
> Mr. Chairman, are the grades under the different headings right through the whole
> of the proposals which have been placed before us for our consideration. In the
> first place, there is the grading under the heading of fitters and turners. So far as
> the two are concerned proposing 46/- and 42/- we are agreed. But we want some
> further discussion on the 36/- and the 34/-. Our opinion is that there is no need to
> have a 34/- man coming under the heading of fitters and turners, inasmuch as it is
> very seldom, or perhaps one could say that it never happens, that a man who has
> never worked upon a machine before is put to do either the most simple fitting or
> to do turning even of the plainest kind. So that we in the first instance want to
> have the 34/- removed, that instead of there being four grades under the heading
> of fitters and turners there shall be three, the 34/- to be moved and the 36/- to be
> made 38/-.
> *Chairman:* You would suggest that a man who would be paid 34/- would be
> called either a filer or a machinist?
> *Mr. Geobey:* That is so . . .
> *Chairman:* Now what was the definition of machinist, do you know, Mr. Barnsley?
> *Mr. Barnsley:* You mean the top man?
> *Chairman:* No, they have got six sections of machine operations on this list. It
> seems to me the driller is essentially a machinist. You see they take machinists
> and split them up for each class of work. Really, you do away with the miller and
> turner, and so on. You have the driller and the machinist: what is the difference?
> *Mr. Geobey:* In the case of the borers a man is never taken straight out of the
> street and put on to boring, therefore there is no need for what can be considered
> a starting rate as we understand it to be. We understand that the 34/- is a rate

which will be paid to a man who has never worked upon a machine before.
Chairman: Yes, a man who goes on as the worst class of machinist.
Mr. Geobey: Taken from the floor as we term it . . .
Chairman: You are rather going away from the whole idea in our minds in coming to this agreement, to form some ladder by which these men can go from the unskilled to the skilled rate.
Mr. Bailey: We are giving it in a great many instances.
Chairman: But you are taking it away in rather the most important ones . . .
Mr. Geobey: Our point, again, Mr. Chairman, is that in the case of machinists and in the case of millers we admit that there is a necessity to have a low rate, because there has to be repetition work within the reach of men who lead what can be called laboring occupations . . .
Chairman: Well, we have discussed these points, Mr. Geobey, and we should be agreeable to leaving out this 34/- in the case of the turners, but we think it rather more important to keep it with the fitters, and also in the borers and the rough filers. But in the other cases where you suggest either the 36/- or the 38/- should be eliminated we think you are making a mistake from your own point of view, because if you take the position of the employer he will consider a great deal longer putting a man up from 34/- to 40/- than putting him up to 36/-, and I think you are going to keep the workers down to the 34/- instead of getting them jumped up to the 40/- as you imagine. This agreement was largely suggested to provide this ladder for men to get up from the position of laborers to that of more skilled men, and we think very strongly you are cutting that ladder away.[73]

What is perhaps most interesting about the exchange, in which both sides compromise a little and eventually reach an agreement, is the fact that such shop floor minutiae as this (and much more not recounted here) has by this time become the subject of argument, negotiation, and formalization at a class level far from the shops where the agreement would be put into effect. Long gone were the days where "master and men" would work out such issues on their own.

Wartime conditions tended to create common interests among employers in the munitions industries, and these interests overshadowed underlying differences between employers. Thus, leaders of the 250-member MEF and leaders of the 250-member Birmingham EEF began to work together during the war and formally merged their two organizations as the war ended. But in the postwar world, the underlying differences between the old members of the MEF—who now referred to themselves as the "Allied Trades" as in "Engineering and Allied Trades"—and the engineering firms began to bubble up, undermining solidarity within the Birmingham EEF.

The source of the conflict between the employers lay in their different workforces, differences which in turn reflected the long-term consequences of different strategies for dealing with the problems of production and labor control. For a variety of reasons, British engineering firms had chosen not to de-skill their shops to anywhere near the extent the Allied Trades employers had done during the previous fifty years, and so most engineering firms remained heavily dependent on skilled labor in the early 1920s.[74] And as Jonathan Zeitlin points out, in such firms, "Managerial innovation instead tended to take the form of cheapening and intensifying skilled labor within existing craft-based work processes through methods such as the multiplication of apprentices, the manipulation of incentive payment systems, systematic overtime, and the promotion of semiskilled workers onto simpler types of machinery."[75] Needless to say, such employer efforts were often resisted by organized labor, particularly the ASE/AEU. One such conflict had resulted in a nationwide lockout of the ASE/AEU in 1898—resulting in a defeat for the union—but by 1920, the AEU had rebuilt itself and was, once again, challenging employers on the question of "managerial rights," particularly with respect to the question of mandatory overtime.

The Allied Trades employers within the Birmingham EEF watched the growing tension between the AEU and the EEF with increasing anxiety. These employers, most of whom hired few skilled engineers, wanted no part of the fight between the engineering firms and the AEU. At the same time, they also wanted considerably more autonomy in dealing with their unions than was possible within the organizational structure of the EEF—a structure which required the national Management Board to approve all local agreements. It was for these reasons that, beginning in the fall of 1921, Clive Kenrick, along with other leaders among the Allied Trades employers, launched a determined effort to carve out a realm of autonomy for the Allied Trades employers within the organization.[76] Toward this end, they convinced the Birmingham EEF to approve their proposed "scheme of re-organization by which the Allied Sections should receive autonomy, and would be able to negotiate matters regarding wages and working conditions apart from the Engineering Section."[77] This scheme was sent on to London for approval by the national Management Board early in 1922, where it was set aside until the EEF settled matters with the AEU on the question of "managerial functions."[78]

The lockout of the AEU by the EEF took place on 11 March 1922, a week after notices were posted.[79] Because by this time the Kenricks employed very few skilled craftsmen, the firm, like most other Allied Trades firms in the district, hardly missed the AEU members, and work proceeded as usual.[80] But soon after the AEU was locked out, the EEF insisted that the Allied Trades unions sign the agreement on managerial rights that the AEU had refused to sign, and after a ballot of the rank-and-file voted against agreeing to the EEF's terms, federated firms were instructed to post lockout notices effective 3 May 1922.[81] The extension of the lockout to the Allied Trades unions met resistance on the part of Allied Trades employers throughout Great Britain, but nowhere was this resistance stronger than in the Birmingham district. At several "emergency meetings" of the Management Board in the days after the lockout extension, it was reported that a number of firms were facing sanction by the EEF for failing to post the notices.[82] Indeed, it must have been embarrassing to Clive Kenrick to be present at the 8 May 1922 meeting of the Management Board when, "It was reported that this member (Archibald Kenrick and Sons Ltd.) had not posted the notice of May 3rd, 1922.

Mr. Clive Kenrick, on behalf of the firm, agreed that the notices should be posted forthwith."[83]

It is worth noting that Clive Kenrick missed several meetings of the Management Board in the weeks and months shortly after this meeting—a most unusual occurrence in his long years of service on the Board. And while we cannot say whether Clive's absence was in protest to the lockout extension, there is no doubt that the Kenricks vigorously protested the action. Here we quote in full a letter, dated 5 May (two days before the above meeting, while the Kenricks were still refusing to post the lockout notices), from the directors of the firm to Sir Allan Smith, Secretary of the Engineering Employer's Federation, London:

Dear Sir,

The crisis that has arisen in negotiations with the Trades Unions on workshop management has made it necessary for us to review our position in the Federation as it is now constituted.

We have followed the present dispute very closely from its inception when the Amalgamated Engineering Union upset the Agreement which you had made with them for overtime, and we have watched the spread of the dispute to other trades with increasing anxiety.

In our opinion it was a grave mistake to challenge the Allied Trades Unions to declare their agreement with the formula for Management which the Engineers had rejected on a ballot. The Principle which has been raised has never been an issue with the bulk of our workpeople and we have no hesitation in saying that we should not have had any trouble on this account if we had not been members of your Federation. There is evidently a sharp divergence of interest and practice between our industries and the Engineering trade, apart from the marked difference between ourselves and the Federation, in our view of the recent negotiations.

Our industries are threatened with a serious stoppage and whatever may be the outcome of the dispute, we can only foresee damage to our trade and a feeling of bitterness amongst our workpeople.

It appears to us that, with the federation as it is now constituted, the difficulty in which we now find ourselves placed is certain to recur, and we want to put it to you quite frankly that it is not satisfactory to us to remain members under the present conditions.

We were satisfied to support our Engineering friends in their dispute with the A.E.U. but we are profoundly dissatisfied that the dispute should have been so gratuitously extended to all the other trades.

We do not wish to embarrass you in your present difficulties and this letter is not so much directed to pointing out our dissatisfaction as to asking you to consider whether you can so amend your Constitution as to make it possible for us to remain members without again incurring the risk of being involved in a dispute which should be confined to Engineers.

It is, of course, within our knowledge that some steps have already been taken to separate the Allied Trades in Birmingham from the Engineering trade within the Federation and we are of the opinion that if the present fusion of the

two is to continue, autonomy must be provided in all matters where a National
Lockout or strike is involved.

Whilst making these observations in respect of the Companies which we
control, we know that we are giving expression to the opinions held by a considerable body of Federated Employers.[84]

This letter gives us an excellent understanding of the directors' sense of the
firm's interests and relationships. There are a few points worth noting. First, from
the directors' point of view, the question of managerial rights is a nonissue in their
works, and, presumably, in the factories of most other Allied Trades employers.
Having long ago stripped their workers of the skills upon which they might base a
claim for workplace control, the employers see no threat from their workers on
this issue. This was a battle these employers had fought, and won, many years
earlier, and so they were not pleased to have to fight it again as a consequence of
their alliance with the engineering employers—for whom the struggle with craftsmen over control of the workshops did remain an issue. What *was* important to the
directors, interestingly enough, was maintaining production and avoiding causing
unnecessary "bitterness" among their workpeople. Their objections to the lockout, then, are quite consistent with Clive's efforts to rationalize labor relations,
minimize conflict, and manage via consent, rather than coercion. Again, having
won the struggle for the shop floor with craft workers many years before, the firm
felt itself, in some sense, "beyond conflict" with its workforce; in fact, in the letter
they suggest that they share an interest with their workers in maintaining production. There is no issue at stake here, for the firm, worth either stopping production
or embittering their workers.

Moreover, what trouble the Kenricks and other Black Country metal trades
employers had experienced in recent years came not from their few skilled workers, but rather from the mass of workers organized by the Workers' Union. As the
strike of 1913 showed, the issues involved not whether the Kenricks had the right
to organize production (this the Workers' Union conceded), but whether there
would be a minimum wage, whether workers would be treated fairly, and whether
the employer should provide protective clothing and other "welfare" provisions.
These were bread-and-butter issues of immediate concern to the "bottom dog."

Mindful of the centrality of these issues, Clive Kenrick had, for some years,
taken the lead within the EEF on negotiations with the WU for base rates of pay.
His leadership here was, no doubt, due to the firm's early experience with the WU
and its district organizers (Beard, Geobey, and Varley) going back to the 1913
strike. Indeed, after a year of negotiation, just prior to the lockout, Clive Kenrick
reported reaching an agreement with the unions on a schedule of base pay rates.
Moreover, as just mentioned, they had even developed the rather advanced notion, for this time, of a fluctuating scale that would vary wage rates up or down
depending on the monthly cost-of-living index published by the Board of Trade.
All this work was for naught, however, when the EEF forced the Kenricks to lock

out workers over, for them, a nonissue, right at the moment when they had reached an accord over what was, for them, the real issue: wages. The EEF wanted the Kenricks to fight a battle the Kenricks had already won, and the EEF was unable to help the Kenricks with the issues that really mattered to them in 1922.

Within a month, the lockout was over as union resistance crumbled and the AEU consented to the principle of managerial control.[85] As intimated in the letter sent earlier from the Kenricks to Sir Allan Smith, the lockout redoubled Clive's efforts to amend the EEF constitution in favor of more freedom for the Allied Trades employers. Thus, in July 1922, Clive Kenrick and a delegation from the Birmingham District were back in London and meeting, once again, with Sir Allan on the question of reorganizing the EEF to accommodate the concerns of the Allied Trades employers.[86] A verbatim transcript remains of this meeting.[87] In it, we find Sir Allan Smith, Secretary of the EEF, and Mr. Maginness, Chairman of the Birmingham delegation, hammering away at the problem of how to distinguish an "Allied Employer" from a full-fledged Engineering Employer. Reading between the lines, it seems clear that Sir Allan would rather not make such a distinction at all, and in the end only agrees to have another meeting. What is of particular interest to us occurs toward the very end of the meeting, when the discussion turns to the recent lockout, and the delegation's fears over whether they can expect again to have their firms shut down over a spat between the AEU and the EEF. At this point, other voices are heard, including, briefly, that of Clive Kenrick. The exchange went as follows:

> *Mr. Maginness:* In what category would you put the question of managerial functions?
>
> *Sir Allan Smith:* In that case when we have got to face managerial functions again, I think we should ask each firm "Do you think it worth while to reserve your right to manage or do you think it worth while for us to lock out?"
>
> *Mr. Maginness:* We do manage in our own works.
>
> *Sir Allan Smith:* In several cases that come before the Central Conference we sometimes doubt whether the employer manages the works or the foreman or the shop stewards.
>
> *Mr. Maginness:* We do not often come to Central Conference.
>
> *Sir Allan Smith:* We have not had the same opportunity of testing your management.
>
> *Mr. Ward:* Some of us have.
>
> *Sir Allan Smith:* The point is what is the intention of the employer in regard to what is involved in it and I should think there would be one answer to that if there is any attempt to move from the employers the right to manage their factories.
>
> *Mr. Kenrick:* If there was any real challenge to it.
>
> *Sir Allan Smith:* We cannot and never will be able to lay down the law unless we can carry the firms with us by convincing them that what we are trying to do is what they want to do. The whole thing depends upon voluntary co-operation and it is our business to find out how that can best be effected. Could you leave it at that?

We do not know the tone of Clive Kenrick's remark, but it smacks of bitterness and sarcasm. The comment expresses, this time face-to-face, the sentiment of the letter sent to Sir Allan Smith earlier in the summer, at the moment when the lockout was extended to the mass of the Kenrick workers by Smith and his national Management Board. What Kenrick is reminding Smith, once again, is that in the Allied Trades firms—Kenrick's hollow-ware works in particular—there has been no serious challenge to the rights of the owners to manage their firms in a generation.

Our story of the social organization of work at the Kenrick firm ends, then, with an irony we might call "the revenge of the craftsmen." Faced with craft worker resistance in the 1890s, the firm replaced these men with less expensive and less troublesome machines and unskilled workers. This strategy worked for a time, but eventually these workers came together, and with the strike of 1913 they forced the firm to recognize minimum wages and bargain with workers collectively. In response, the Kenricks joined with other beleaguered employers to form the MEF to protect themselves against the union. But after the war, with the merger of the MEF and the EEF, the firm's fate was, once again, tied to a struggle with craft workers. And in the lockout of the AEU by the EEF in 1922, we find the firm in the absurd position of fighting a battle against craftsmen they believed they had won many years earlier. In this sense, their "class interests," as then constituted and defined, worked directly against their interests as a firm of hollow-ware makers.

Nevertheless, having thrown in with the Birmingham engineering employers, the firm would not soon escape the responsibilities, as well as the benefits, of coordination with their capitalist class comrades. Jumping ahead a few years to the aftermath of the general strike of 1926, we find Clive Kenrick at a meeting of the Management Board listening to the following letter written by the Chairman of the Birmingham EEF to the Management Board:

> I think it is very desirable that at the Board Meeting on Tuesday next we should take an opportunity of reviewing the situation which exists as a result of the collapse of the General Strike. Apart from other considerations, the question in varying phases is being raised by members of the Association.
>
> It seems to me at the moment we are in an extraordinarily interesting situation, fraught with great possibility. For twenty years, trade unionists have been in possession of a machine which they have used defensively and offensively against the employers. For the past ten years at least, the aggregation of the unions in the TUC has made possible the threat of a general strike, the menace of which has affected not only industrial but social and political policy alike.
>
> The result undoubtedly was, in my opinion, that the workers developed a sense of power and independence which, unfortunately, they used, not in the direction of cooperation, but rather in promoting what they conceived to be a class interest. There was a certain feeling that, when they so cared, they could put the 'boss' into his place; that Jack was in fact better than his master. The effect on

the workers psychology has not been good from the point of view of efficiency of production. There has been no community of interest and little sense of loyalty.

What is the position today? The machine upon which the men relied has utterly failed them. I think they now realize that the community is greater than Trade Unions. The prestige of the Unions has been shattered to a degree that must undoubtedly be reflected in the membership. The credit of the leaders is at an enormous discount. In other words, their standards have been destroyed, and in one fashion or another, have to be rebuilt.

The question arises as to what is to be done with such a situation. Are the employers to sit still until the Unions have time to recover from the blow and rehabilitate themselves, or are they capable of developing a constructive policy that will lead working-class tendencies towards a sense of unity with employers and towards a sense of necessity for making a combined effort to achieve an efficient industrial machine? I think that at this stage the matter is certainly worth some discussing and it would be extremely interesting to know your views on the subject.[88]

At the Chairman's behest, Clive was appointed to a committee to make recommendations to the Management Board. This committee reported back to the Board on 6 July1926, with a series of recommendations designed "for ensuring that workpeople have a personal interest in the prosperity of their Employers." These included:

1. Share capital
2. Deposit fund with profit-sharing rights
3. Hypothetical shares
4. Piecework system
5. "Educational propaganda"
6. Social welfare
7. Youth employment
8. Benevolent funds
9. Sports teams[89]

Thus, much as the activities of the unions in 1913 had presented the employers with both problems and opportunities, so too the near collapse of the unions thirteen years later confronted the employers with similar challenges. In this instance, the opportunity was to reassert *bourgeois* dominance; the problem was to legitimate it. Thus, in their proposals, we find them offering neopaternalist strategies reminiscent of the mid-nineteenth century, strategies that, for a time, had captured the hearts, minds, and labor power of workers in the British metal trades.

Shaping Struggles: Bureaucratic Hegemony, 1914-1922

In this final episode of our story, we see yet another distinctive shift in factory politics, this time toward, what we call, "Bureaucratic Hegemony." With this new type of administration, workplace politics were increasingly embedded within a bureaucratic framework where wages, labor relations, the division of labor, and work performance were set out within the formal rules of rational, legal authority and governed increasingly by *supra*-firm organizations. In this sense, we saw how industrial conflict was *institutionalized*—along with an unequal sex-caste system—rather than *repressed*.[90] As Scott Davies characterizes this trend, "Institutions such as unions, grievance procedures, collective bargaining, and universal rules of procedure create norms of 'industrial justice' that must be followed by both unions and management . . . Collective bargaining thus represents the institutionalization of a created common interest between capital and labour, resting on a common precondition of profitability."[91]

The demise of paternalism and the resulting class conflict brought about both capitalist and working class formation and collective action. On the one hand, we saw owners colluding within the tightly knit Midland Employers' Federation that was formed as a direct response to the confrontations of 1913 and the growing influence of the Workers' Union. This cooperation, along with previous agreements among the CHIMA members and the shared bounty of lucrative war contracts, helped keep interfirm competition in check. On the other hand, while the labor process at Kenricks had been for some time mechanized and deskilled, workers were now less dependent on and vulnerable to their employers owing to the power of their own labor organization and collective action, a high demand for their labor during the war, and state intervention in the form of minimum wage laws, social insurance schemes, and labor legislation.

Unlike the two previous regimes that were founded on forms of patriarchal despotism or a "stern yet considerate paternalism," these changing circumstances meant that workers now had to be, to some extent, *persuaded* to cooperate with management, and their interests, as much as possible, had to be coordinated with those of capital. This new regime was dependent, then, on a degree of voluntary "consent" by workers to both managerial prerogative and to unequal economic relations. We believe that a degree of worker consent was produced in the Midlands metal trades, and at the Kenrick factory in particular, to the extent that labor believed that "industrial justice" was real and that their union and the "machinery of negotiation" offered them at least a modicum of recognition and respect. Moreover, with the significant shift to a female work force, as well as the demands of the "war effort," sex-specific "welfare supervision" techniques and "protection" legislation were introduced as a means to garnish female worker cooperation and to enhance discipline and productivity in the shops. These state-initiated tactics represent, as Mary Lynn Stewart suggests, a form of welfare state "social patriarchy" that was aimed at "protecting" women workers while, at the same time, appeasing their penchant to down tools.[92] As some have suggested, these early welfare state policies, while apparently beneficial to women, actually reinforced their dependence on the patriarchal family.[93] Clearly, the role of the state and the wartime emergency was pivotal in shaping the nature of this bureaucratic factory regime.

As we have shown, a bureaucratic regime subordinates *both* workers and owners to a framework of negotiation and the impersonal rule of law. Observing developments in Birmingham during the period, Dennis Smith writes, "as management became more bureaucratic it found that stability was enhanced by an expansion of the role of organized trade unions. Paternalism and craft solidarity yielded ground to organizational rationality."[94] Although signing on with the WU appeared to advance the interests of Kenrick workers, offering them hard-won wage concessions and other benefits, the union also tended to suppress working class militancy and to channel their energies in a reformist direction. And, as we have shown, strategies of incorporation and collusion between the union and the owners helped institutionalize a system of sex-based wage differential that would later affect the majority of the Kenrick workforce—60 percent of whom were women by 1918, 75 percent of those over the age of eighteen. Moreover, it was only by entering into agreements with the unions—a form of managerial "consent" to the power of organized labor—that the Kenricks and other Midlands employers were able to secure a reasonable degree of cooperation from their workforce and put a lid on militant labor action. Yet, the Kenricks paid dearly for their concessions. By submitting themselves to the machinery of negotiation, the employers lost a great deal of their sovereignty as businessmen, turning themselves, as we put it, from patriarchal autocrats into bureaucrats. Not only were they now forced to "negotiate" an appreciable amount of shop floor minutiae with union representatives, but their amalgamation with other owners bridled their ability to act as independent proprietors. The extent of this subordination was most dramatically displayed when they found themselves, as they did in 1922, in the seemingly senseless position of having to stand together in solidarity with other members of the EEF to lock out their own workers despite not having any grievance with them. In sum, in our analysis the factory regime of "bureaucratic hegemony," we highlighted the importance of bureaucratized, "industrial relations," and the influence of more extensive state intervention. Combined, these two forces reduced labor's dependency on employers and helped check managerial despotism, but also stifled working class radicalism.

Chapter 6

Epilogue

In this book, we have examined three distinct political regimes of production at the Kenrick factory between the years of 1791 and 1922, and we have argued that they represented successive forms of capitalist patriarchy. Indeed, as we have shown, the Kenricks were able to maintain their *power* and their *profits*, to a great extent, because they were able to use *patriarchy*, or, as Charles Tilly suggests, "categorical distinctions"—in this case, between men and women and between children and adults—to solve "pressing organizational problems." Patriarchy, thus, seems to us to have been an essential social resource—a form of social and cultural capital—that the Kenricks were able to convert into private wealth. These factory regimes mediated, regulated, and shaped struggles between workers and owners/managers, and, in turn, shaped the collective interest and action of these parties.

The first regime was characterized by despotism and a form of patriarchy that was imported into the early factory from an earlier "proto-industrial" period. The patriarchal authority of laboring subcontractors regulated and reproduced the social relations of production, and once it was undermined by a series of social and economic contradictions, it gave way to a second, more paternalistic regime. We see this form of despotic paternalism as a legitimating ideology for the inequality and instability that was produced by the owner's business strategies. This paternalism, we contend, helped contribute to the less militant stance of skilled male workers and thereby shaped struggles over mechanization and the substitution of female labor in favor of capital.

In the end, the forces and conditions existent under this paternalistic regime proved instrumental in the demise of the system of subcontracting and its associated handicraft production. Specifically, the social organization of work in the more traditional form of adult male subcontracting was eventually replaced by direct control of female piece-rate machine "operatives" by an increasing number of male company foremen. Yet, by creating these "really" subordinated workers, the Kenricks themselves set the stage for working class resistance, particularly from their female workers. By refusing to extend their paternalism to these "unskilled" female workers, this ideology was less effective at binding the workforce to the workplace. Moreover, by failing to organize these women into a loyal workforce, the owners left themselves vulnerable to having these marginalized and increasingly disaffected women workers organize against them. This vulnerability was exploited in April of 1913 when the Workers' Union (WU) organized unskilled Kenrick workers in a dramatic two-week strike for union recognition and a scale of minimum wages. But the strike did not result in a political economic meltdown because patriarchy provided the point of agreement, or consensus, around which working class men and their employers could work out their (class) differences, resulting in both the preservation of capitalism and the reassertion of male authority.

Finally, although the replacement of skilled workers with unskilled workers was initially profitable, it became clear—dramatically so during the unrest of 1913—that company paternalism, as an effective political tool, was through. A new form of factory regulation was needed, and we saw a third regime emerge. Widespread working class impoverishment and unrest, as well as the demands of war, brought about state intervention and regulation. The state's role was to stabilize the social formation by bringing organized capital and labor to the bargaining table. Part of that strategy was to provide the working class with some minimal level of social and economic security, and to discourage strikes and social unrest by encouraging the peaceful settlement of industrial conflict through "conciliation." As we have suggested, the goal of state intervention was to offer a more level playing field and provide new rules for the game played on it. These forms of social insurance and *re*assurance altered the dependency of working class people on wage labor and put some limits on arbitrary despotism. Under this "hegemonic" regime, indicative of this stage of capitalism, not only is "the application of coercion circumscribed and regularized, but the infliction of discipline and punishment itself becomes the object of consent."[1] As we have shown, factory politics and interests—for both labor and capital—were then embedded within a bureaucratic regime set within the formal rules of rational and legal authority. This regime, while in some ways empowering workers, also seemed to channel any of their remaining dissatisfaction in a reformist direction.

A New World Order?

If state intervention in Britain at the turn of the twentieth century appeared to hold out the promise of "industrial peace," the country's present neoliberal political regime, by contrast, appears determined to turn back the clock on "regulated capitalism." Indeed, today, England is deeply mired, at the close of that same century, in the challenges of the new global economy. Economic "globalization," or a process of transnational market expansion and integration, is characterized by "a new international division of labor, the global spread of financial markets, an interpenetration of industries across borders, the spacial reorganization of production, a temporal acceleration in economic activity, vast movements of populations, a diffusion of consumer goods, and a welter of transnational cultural linkages."[2] While certainly not a new phenomenon, this phase of economic globalization is being facilitated—albeit unevenly and, in some instances, only partially—by nation-states throughout the metropolitan world pursuing so-called "third way" (neither traditionally "left" nor "right") political agendas. This has been the hallmark of Anglo-American politics in the era of Tony Blair and Bill Clinton.

Under the threat of "global competition" and the demands of transnational corporations (TNCs), Britain, following the U.S. lead, has pursued policies such as deregulation, privatization, and "free" trade and has facilitated the infrastructure for global commodity, capital, and stock markets, and telecommunications systems. For states like Britain, these strategies appear to enhance their international stature as leaders in the new cosmopolitan world system and, to some extent, to capitalize on their short-term political and economic opportunities. Yet, these policy measures—taken in combination with neoliberal social agendas and rollbacks in the welfare state—only serve to cast such nation-states as *both* facilitators and potential victims of globalization. This contradictory position comes about, in part, according to William Sites, because the state's relationship with the process of globalization engenders "the 'separation' of social actors—corporations, citizens, residents—from the sociopolitical conditions and spacial patterns that anchored an earlier era . . . [conditions and patterns that] not only supported national economic activities but also rooted them, and social actors generally, in certain patterns of civic life."[3] The so-called "flexibility" of production and capital mobility that highlight globalization are, as some analysts contend, helping TNCs to weaken unions, cut costs, reduce wages, escape state regulatory legislation and, therefore, to increase profits. In short, state managers have empowered the position of the TNCs at their own expense—to say nothing of the position of the working classes—impeding their ability to govern their own countries and maintain their own legitimacy. A recent episode, close to the center of our story, is illustrative.

In the spring of the year 2000, people came from all over Britain to the streets of Birmingham, as one newspaper put it, "angry, defiant, and feeling betrayed." In the biggest labor protest in decades, 80,000 people marched to rebuke a possible death sentence for Britain's auto industry. The action followed the announcement that German automaker BMW intended to close, rather than reinvest in and strengthen, its Rover subsidiary plant. According to the paper, people packed the streets to angrily denounce the sell-out of Rover's Longbridge plant and the possible loss of some 6,000 jobs. Even as the Midlands sky filled with rain and temperatures dropped, the enthusiasm of the two-mile-long protest was not dampened. Thousands not involved in the rally lined the streets to cheer the marchers on.[4] One group held an effigy of Prime Minister Blair dressed in a blue BMW suit chanting, "Blair doesn't care." Much like the Kenrick factory in the nearby hamlet of West Bromwich a hundred years before, in contemporary Birmingham, it is said, everyone knows someone who works at Longbridge. Undoubtedly, a few of those 6,000 jobs belonged to the descendants of Kenrick workers. Forewarning the BMW-Rover debacle, Michael Burawoy suggested nearly twenty years ago that, "Irrespective of state interventions there are signs that in all advanced capitalist societies that hegemonic regimes are developing a despotic face. . . . In this period one can anticipate the working classes beginning to feel their collective impotence and the irreconcilability of their interests with the development of capitalism, understood as an international phenomenon."[5]

Clearly, for many of the threatened workers at Longbridge, the "despotic face" of the new regime is not only that of BMW, but that of Tony Blair and the policies of his neoliberal, "New Labor" government. Responding to the protests in Birmingham, Mr. Blair wrote, in typical managerialist fashion, "If governments in the past, of both major political parties, have been drawn towards 'rescuing' a company in difficulties, we see our role now as helping to equip people and businesses for the new economy."[6] He insisted the government was not leaving workers in the lurch but saw its role as "managing transition" rather than intervening to bail out businesses. We see then, in this emerging regime of hegemonic despotism, an attempt to return, yet again, to the laissez-faire policies and practices of the early nineteenth century, only this time on a global scale. But as Mark Weisbrot reminds us:

> It has long been known that a system of unregulated markets does not regulate itself, is prone to crises and even depressions, and does not necessarily allow the majority of its participants to share in the gains from economic growth and technological progress. Globalization is a way of forgetting all this, of blotting out the last two centuries of economic history as though it were all part of a bad dream. It is capitalism in denial, a back door way of re-introducing the worst excesses and irrationalities of the market, long ameliorated, to varying degrees in different countries, by the nation-state.[7]

Is the nation-state withering away as some "hyperglobalists"contend? Hardly. National governments and nationalism persist as potent and determining features of our age. What remains to be seen is whether nation-states pursuing neoliberal policy agendas can possibly establish the kind of "durable political and social conditions that will be demanded over the long term by global/national and global/local linkages posed by international economic integration."[8] History suggests that they cannot.

If economic globalization is enhancing the *power* and the *profits* of TNCs, then what role does *patriarchy* play in the current phase of the world economy? As our historical portrait has shown, in the context of preexisting patriarchal relationships, the ways in which women experienced *industrialization* permitted men to secure a better deal out of the social and economic changes taking place. The available evidence suggests that women's experience of *globalization* is correlative. A recent UN study reports that: Women now comprise an increasing share of the world's labor force—at least one-third in all regions except northern Africa and western Asia. Self-employment and part-time and home-based work have expanded opportunities for women's labor force participation, but these opportunities are characterized by lack of security, lack of benefits, and low income. The informal sector is a larger source of employment for women than for men. More women than before are in the labor force throughout their reproductive years, though obstacles to combining family responsibilities with employment persist. Women, especially younger women, experience more unemployment than men and for a longer period of time than men. Women remain at the lower end of a segregated labor market and continue to be concentrated in a few occupations, to hold positions of little or no authority, and to receive less pay than men. Estimates are that today, while women do two-thirds of the world's work, they earn only one-tenth of the world's income and own only one-hundredth of the world's property.[9] Overall, women and children are disproportionately the world's poorest citizens, and, as Susan H. Williams contends, the process of globalization relies on and often exaggerates the gendered division of labor that is used to justify and perpetuate this inequality. In other words, gender and poverty are related but independent reasons for the particularly harsh impact of economic globalization on women."[10]

Coming full circle then, we see that, much like the Kenricks during the nineteenth and early twentieth centuries, contemporary TNCs have been able to benefit from and reproduce the unequal power and authority relations of class, gender, and age. Yet, like the other regimes of early capitalism analyzed here, the complex, historically contingent, and multifaceted process of contemporary globalization is rife with contradictory fissures. In the streets of Birmingham and in other parts of the globe, people are increasingly aware of their common plight in the face of global capitalism. The same interconnectedness that permits patterns of inequality and hierarchy to flourish may also help create networks for new transnational and transcultural movements of solidarity and resistance. Such a global network of participation, opposition, and reform may hold out the hope of trading long-lasting "durable inequalities" for a more truly democratic future.

Notes

Chapter 1

1. Ironically, the Kenrick family lost control of the firm in 1991, their bicentennial year, when the firm was bought by another Birmingham metal fabricator.

2. Michael Burawoy, *The Politics of Production: Factory Regimes under Capitalism and Socialism* (London: Verso, 1985), "Karl Marx and the Satanic Mills: Factory Politics under Early Capitalism in England, the United States, and Russia," *American Journal of Sociology* 90, no. 2 (September 1984): 247-282, and "Between the Labor Process and the State: The Changing Face of Factory Regimes under Advanced Capitalism," *American Sociological Review* 48, no. 5 (October 1983): 587-605.

3. Charles Tilly, *Durable Inequality* (Berkeley: University of California Press, 1998), 7-8.

4. Heidi Hartmann, "Capitalism, Patriarchy, and Job Segregation by Sex," in *Women and the Workplace: The Implications of Occupational Segregation*, eds. Martha Blaxall and Barbara Reagan (Chicago: University of Chicago Press, 1976), 168. See also, Heidi Hartmann, "The Unhappy Marriage of Marxism and Feminism: Towards a More Progressive Union," *Capital and Class* 8 (summer 1979): 1-33; Ruth Milkman, "Organizing the Sexual Division of Labor: Historical Perspectives on 'Women's Work' and the American Labor Movement," *Socialist Review* (January-February 1980): 105-108; Judy Lown, "Not So Much a Factory,

More a Form of Patriarchy: Gender and Class during Industrialization," in *Gender, Class and Work*, ed. Eva Gamarnikow, David H. J. Morgan, June Purvis, and Daphne E. Talorson (London: Heinemann, 1983), 28-45, and her *Women and Industrialization: Gender at Work in Nineteenth Century England* (Minneapolis: University of Minnesota Press, 1990). Similarly, Veronica Beechey writes "Capitalism did not create the patriarchal family but developed on the basis of the patriarchal domestic economy which was already in existence." See her "On Patriarchy," *Feminist Review* 3 (1979): 66-83.

5.   Many valuable studies of women workers under capitalism have been written. For a review, see Laura Frader and Sonya O. Rose, eds., *Gender and Class in Modern Europe* (Ithaca: Cornell University Press, 1996). See also "Bibliography: Women and Work," *Journal of Women's History* 1 (spring 1980): 138-169; Ava Baron, ed., *Work Engendered: Toward a New History of American Labor* (Ithaca: Cornell University Press, 1991); Eileen Boris, "Beyond Dichotomy: Recent Books in North American Women's Labour History," *Journal of Women's History* 4, no. 3 (winter 1993): 162-179; and the work that has been done on women workers in England during the late nineteenth and early twentieth centuries, including Jane Lewis, ed., *Labour and Love: Women's Experience of Home and Family, 1850-1940* (Oxford: Oxford University Press, 1985); Angela V. John, ed., *Unequal Opportunities: Women's Employment in England 1800-1918* (Oxford: Basil Blackwell, 1985); Sonya O. Rose, "'Gender at Work': Sex, Class and Industrial Capitalism," *History Workshop Journal: A Journal of Socialist Historians* 21 (1986): 113-131, "Gender Segregation in the Transition to the Factory: The English Hosiery Industry, 1850-1910," *Feminist Studies* 13, no. 1 (spring 1987), "Gender Antagonism and Class Conflict: Exclusionary Strategies of Male Trade Unionists in Nineteenth-Century Britain," *Social History* 13, no. 2 (1988): 191-208, and *Limited Livelihoods: Gender and Class in Nineteenth-Century England* (Berkeley: University of California Press, 1992); Jeffery Haydu, *Between Craft and Class: Skilled Workers and Factory Politics in the United States and Britain, 1890-1922* (Berkeley: University of California Press, 1988); Laura Lee Downs, *Manufacturing Inequality: Gender Division in the French and British Metalworking Industries, 1914-1939* (Ithaca: Cornell University Press, 1995). While important, these studies of women workers are organized around occupation, union, industry, community, or time period. And, while some inevitably include detailed data on, and analysis of, the work place, historical case studies of individual firms are rare, see Lown, *Women and Industrialization*; Deborah Thom, "Women at the Woolwich Arsenal 1915-1919," *Oral History* 6, no. 2 (1978): 58-73. For Canada, see Joy Parr, *The Gender of Breadwinners: Women, Men and Change in Two Industrial Towns, 1880-1950* (Toronto: Toronto University Press, 1990) and for the United States, see Carole Turbin, *Working Women of Collar City: Gender and Class and Community in Troy, New York, 1864-86* (Urbana: University of Illinois Press, 1992).

6. Patrick Joyce, "The Historical Meanings of Work: An Introduction," in *The Historical Meanings of Work*, ed. Patrick Joyce (Cambridge: Cambridge University Press, 1987), 10.

7. Lown, "Not So Much a Factory, More a Form of Patriarchy," 36.

8. See note 5.

9. Harry Braverman, *Labor and Monopoly Capital: The Degradation of Work in the Twentieth Century* (New York: Monthly Review Press, 1975).

10. Implied, as well, was the Marxist assumption that capitalist social relations are inherently antagonistic and that class formation and collective action necessarily spring from these structural positions. See Karl Marx and Frederich Engels, *The Communist Manifesto: A Modern Edition* (London: Verso, 1998 [1848]).

11. Richard Edwards, *Contested Terrain: The Transformation of the Workplace in the Twentieth Century* (New York: Basic Books, 1979); Andrew L. Friedman, *Industry and Labour: Class Struggle at Work and Monopoly Capitalism* (London: Macmillan, 1977); Craig Littler, *The Development of the Labour Process in Capitalist Societies: A Comparative Study of the Transformation of Work Organization in Britain, Japan, and the USA* (London: Heinemann, 1982).

12. Michael Burawoy, *Manufacturing Consent* (Chicago: University of Chicago Press, 1979).

13. In particular, see David M. Gordon, Richard Edwards, and Michael Reich, *Segmented Workers, Divided Workers* (Cambridge: Cambridge University Press, 1982).

14. David Noble, *American by Design* (Oxford: Oxford University Press, 1977), and "Social Choice in Machine Design: The Case of Automatically Controlled Machine Tools," in *Case Studies in the Labor Process*, ed. Andrew Zimbalist (New York: Monthly Review Press, 1979), 18-50; Michael Wallace and Arnie Kallenberg, "Industrial Transformation and the Decline of Craft," *American Sociological Review* 47, no. 3 (1982): 307-324; Andrew Zimbalist, "Technology and the Labor Process in the Printing Industry," in *Case Studies in the Labor Process*, ed. Andrew Zimbalist (New York: Monthly Review Press, 1979), 103-126.

15. Joyce, *The Historical Meanings of Work*, 6.

16. See Burawoy, *The Politics of Production*.

17. Geoff Eley, "Edward Thompson, Social History and Political Culture: The Making of a Working-Class Public, 1780-1850," in *E. P. Thompson: Critical Perspectives*, eds. Harvey J. Kaye and Keith McClelland (Philadelphia: Temple University Press, 1990), 13.

18. E. P. Thompson, *The Making of the English Working Class* (New York: Vintage Books, 1963).

19. Thompson, *The Making of the English Working Class*, 9.

20. See, for example, Michael Hanagan, *The Logic of Solidarity: Artisans and Industrial Workers in Three French Towns, 1871-1914* (Urbana: University of Illinois Press, 1980); Joan Wallach Scott, *The Glass Workers of Carmaux* (Cam-

bridge, Mass: Harvard University Press, 1974); Patrick Joyce, *Work, Society and Politics: The Culture of the Factory in Later Victorian England* (New Brunswick: Rutgers University Press, 1980); Victoria Bonnell, *Roots of Rebellion: Workers' Politics and Organizations in St. Petersburg and Moscow, 1900-1914* (Berkeley: University of California Press, 1983); Ira Katznelson and Aristide Zolberg, eds., *Working Class Formation: Nineteenth Century Patterns in Western Europe and the United States* (Princeton: Princeton University Press, 1986).

21. Frader and Rose, "Introduction," in Frader and Rose, *Gender and Class in Modern Europe*, 2.

22. For example, Lenard Berlanstein declared that "the new labor history has quickly become the 'old new labor history.'" See Lenard R. Berlanstein, ed. *Rethinking Labor History: Essays on Discourse and Class Analysis* (Urbana: University of Illinois Press, 1993), 4.

23. For example, ten years earlier, Gareth Stedman Jones radically changed direction from his early Marxist orientation by claiming that Chartism was not a response to the material conditions of capitalism but, instead, was a product of political discourse. See his *Languages of Class: Studies in English Working-Class History, 1832-1982* (Cambridge: Cambridge University Press, 1983).

24. William H. Sewell Jr. "Toward a Post-Materialist Rhetoric for Labor History," in Berlanstein, *Rethinking Labor History*, 18.

25. Jane Gray, "Gender and Uneven Working-Class Formation in the Irish Linen Industry," in Frader and Rose, *Gender and Class in Modern Europe*, 48.

26. Ava Baron, "Gender and Labor History: Learning from the Past, Looking to the Future," in Baron, *Work Engendered*, 17.

27. Rose, *Limited Livelihoods*, 13-14.

28. Lown, "Not So Much a Factory, More a Form of Patriarchy," 33.

29. Rose, *Limited Livelihoods*, 19.

30. Researchers who have attempted to include gender in Burawoy's framework include: Scott Davies, "Inserting Gender into Burawoy's Theory of the Labour Process," *Work, Employment & Society* 4, no. 3 (September 1990): 391-406; David Knights and Hugh Willmott, "Power and Subjectivity at Work: From Degradation to Subjugation in Social Relations," *Sociology* 23, no. 4 (1989): 535-558; C. Gannage, "A World of Difference: The Case of Women Workers in a Canadien Garment Factory," in *Feminism and Political Economy: Women's Work, Women's Struggles*, eds. H. Luxton and M. Maroney (Toronto: Meuthuen, 1987); Craig Heron and Robert H. Storey, *On the Job in Canada* (Kingston: McGill-Queens University Press, 1986); Ching Kwan Lee, "Familial Hegemony: Gender and Production Politics on Hong Kong's Electronic Shopfloor," *Gender and Society* 7, no. 4 (1993): 529-547, and "Engendering the Worlds of Labor: Women Workers, Labor Markets, and Production Politics in the South China Economic Miracle," *American Sociological Review* 60, no. 3 (June 1995): 378-397. This critique of Burawoy's perspective is valid and applies as well to the work inspired by Burawoy, including William G. Staples, "Technology, Control, and the Social Organization

of Work at a British Metal Trades Firm, 1791-1891," *American Journal of Sociology* 93, no. 1 (1987): 62-88. Until Lee's comparative ethnography of gendered production regimes in two south China electronics factories, however, critics of Burawoy's work on production politics had not gone much beyond pointing out that gender is missing from his theoretical framework.

31. "Anarchy in the market leads to despotism in production: the market is constitutive of the apparatus of production," Burawoy, "Karl Marx and the Satanic Mills," 251.

32. Karl Marx, *Capital*, vol. 1 (New York: Vintage Books, 1977 [1867]), 645, 1019-1038.

33. Burawoy, "Between the Labor Process and the State," 589.

34. Burawoy, "Between the Labor Process and the State," 590, italics in the original.

35. Alan Warde, "Industrial Discipline: Factory Regime and Politics in Lancaster," *Work, Employment & Society* 3, no. 1 (1989): 51.

36. See, for example, Craig Littler and G. Salaman, "Bravermania and Beyond: Recent Theories of the Labour Process," *Sociology* 16, no. 2 (May 1982): 251-269; M. Storper and R. Walker, "Theory of Labour and the Theory of Location," *International Journal of Urban and Regional Research* 7, no. 1 (1983): 1-44.

37. Rose and Frader, "Introduction," *Gender and Class in Modern Europe*, 32.

38. Anthony Giddens, *Central Problems in Social Theory* (Berkeley: University of California Press, 1983).

39. As Judy Lown points out, paternalism is an *outcome* of the stratification system and not the source of it. See Lown, "Not So Much a Factory, More a Form of Patriarchy," 35.

40. Charles Tilly, *Durable Inequality*, 7-8.

41. Burawoy, "Between the Labor Process and the State," 590.

Chapter 2

1. Archibald Kenrick I diary, 15 February 1787, cited in R. A. Church, *Kenricks in Hardware: A Family Business, 1791-1966* (Newton Abbot, England: David and Charles, 1969), 23-24. He again wrote that, "I told them in the afternoon that I would not have any loss of time, that if they neglected my business they might go to those who would put up with it, that I was determined I would not let them be good for nothing . . ." Archibald Kenrick I diary, 7 February 1787, cited in Church, *Kenricks in Hardware*, 23. Max Weber stated, "A man does not 'by nature' wish to earn more and more money, but simply to live as he is accustomed to live and to earn as much as is necessary for that purpose. Wherever modern capitalism has begun its work of increasing the productivity of human labour by increasing its intensity, it has encountered the immensely stubborn resistance of this leading trait of pre-capitalistic labour." Max Weber, *The Protestant Ethic and the Spirit of Capitalism* (New York: Scribner and Sons, 1958), 60.

2. Archibald Kenrick I diary, 1787, cited in Church, *Kenricks in Hardware*, 21. Weber again is pertinent: "A flood of mistrust, sometimes of hatred, above all of moral indignation, regularly opposed itself to the first innovator. . . . It is very easy not to recognize that only an unusually strong character could save an entrepreneur of this new type from the loss of his temperate self-control and from both moral and economic shipwreck." Weber, *The Protestant Ethic*, 69.

3. For a discussion of manufacturers in the district see Clive Behagg, *Politics and Production in the Early Nineteenth Century* (London: Routledge, 1990).

4. G. C. Allen, *The Industrial Development of Birmingham and the Black Country, 1860-1927* (London: Allen and Unwin, 1929), 7.

5. See the detailed and descriptive study of the metalworking families of the region by Marie B. Rowlands, *Masters and Men in the West Midland Metalware Trades before the Industrial Revolution* (Manchester: Manchester University Press, 1975), 7. A "copyholder" held rights to land on an estate as evidenced by a written document in the records of a manorial court. This was less than a "freeholder," in that the latter had the right to pass on his property through inheritance.

6. Allen, *The Industrial Development of Birmingham*, 84-98; W. H. B. Court, *The Rise of the Midland Industries, 1600-1838*, 2nd ed. (Oxford University Press, 1953 [1938]), 78-99; Rowlands, *Masters and Men*, 54-65.

7. Conrad Gill and Asa Briggs, *History of Birmingham* (London: Oxford University Press, 1952), 88-89.

8. Rowlands, *Masters and Men*, 43. See also P. M. Frost, "The Growth and Localization of Rural Industry in South Staffordshire, 1560-1720" (Ph.D. diss., University of Birmingham, 1973), 278-350.

9. Rowlands, *Masters and Men*, 5, 43; Court, *Midland Industries*, 22.

10. Rowlands, *Masters and Men*, 1.

11. Rowlands, *Masters and Men*, 20. For histories of West Bromwich see, Richard D. Woodall, *West Bromwich Yesterdays: A Short Historical Study of "The*

City of a Hundred Trades" (Sutton, England: Norman A. Tector, 1958); D. Dilworth, *West Bromwich before the Industrial Revolution* (Tipton, England: Black Country Society, 1973).

12. Franklin Mendels, "Proto-Industrialization: The First Phase of the Industrialization Process," *Journal of Economic History*, 32, no.1 (1972): 253; see also Rowlands, *Masters and Men*, 39-53; Peter Kriedte, Hans Medick, and Jurgen Schlumbohm, eds., *Industrialization before Industrialization: Rural Industry in the Genesis of Capitalism* (Cambridge: Cambridge University Press, 1981), 38-73.

13. Concerning women and children, Rowlands writes, "Wills show father and sons working in the same workshops. It is not clear whether women and children worked . . . Casual labor by women and children in the workshops would have been natural and inevitable in the family and the extent of this participation would depend a great deal on the individuals concerned," Rowlands, *Masters and Men*, 39.

14. Maxine Berg, "Women's Work, Mechanization, and the Early Phases of Industrialization in England," in *The Historical Meanings of Work*, ed. Patrick Joyce (Cambridge: Cambridge University Press, 1987), 76.

15. Berg, "Women's Work, Mechanization, and the Early Phases of Industrialization in England," 83-88, and Appendices.

16. Rowlands, *Masters and Men*, 78.

17. Rowlands, *Masters and Men*, 80. See also Kriedte Medick, and Schlumbohm, *Industrialization before Industrialization*,136-138.

18. And despite the reputed poverty among these domestic workers, particularly that of the nailers, Rowlands concludes after her examination of their estates that both their houses and furnishings did not differ significantly from the houses of comparable persons in other parts of the Midlands or the country for that matter. She states that, "The evidence of probate records reveals the metalworkers as a group of people with many different levels of skill and of many different levels of social success. While no doubt this evidence concerns the more prosperous among the tradesmen it does emphasize that a considerable degree of comfort and a margin of profit were attainable," Rowlands, *Masters and Men*, 51.

19. Rowlands, *Masters and Men*, 80.

20. Sometimes the two were combined. In 1802, an Act of Parliament permitted the enclosure of more then 350 acres of open common and pasture within the parish of West Bromwich. Housing was quickly built upon much of the land while several open coal pits were dug in what is today the center of the town. See Woodall, *West Bromwich Yesterdays*, 16.

21. Rowlands, *Masters and Men*, 39-53.

22. In the case of Birmingham, the city's notorious political structure had provided additional incentive. As Gill and Briggs put it, "It has often been suggested that Birmingham and its industries flourished . . . because the town, not being an old chartered borough, was doubly free: there were no gilds to cramp and

discourage enterprise; and it was not subject to the religious trammels imposed by the Clarendon Code, so that there was a strong inducement for nonconformists to settle there. As Hutton said: 'A town without a charter is a town without a shackle.'" Gill and Briggs, *History of Birmingham*, 59.

23. Dennis Smith states, "Ironically, the appearance of unprecedented concentrations of capital and labour in the industrial towns placed an intolerable strain upon the very social system that had made this development possible. Industrialization and urbanization stretched beyond its limit the capacity for political management of institutions adapted to agrarian capitalism, small-scale craft production, and local market trading." See Dennis Smith, *Conflict and Compromise: Class Formation in English Society, 1830-1914: A Comparative Study of Birmingham and Sheffield* (London: Routledge and Kegan Paul, 1982), 7.

24. Rowlands, *Masters and Men*, 165.

25. Rowlands, *Masters and Men*, 157.

26. Rowlands, *Masters and Men*, 157-158.

27. As Marx expressed it, "the fact is that capital subsumes the labour process as it finds it, that is to say, it takes over an *existing labour process.* . . . If changes occur in these traditionally established *labour processes* after their takeover by capital, these are nothing but the gradual consequences of that subsumption." Italics in the original, Marx, *Capital* 1:1021.

28. Stock Book, 1814, Kenrick Collection; Buildings, Tools, and Fixtures Book, n.d., Kenrick Collection, the Black Country Museum, Tipton Road, Dudley DY1 4SQ. Hereafter cited as "Kenrick Collection." When we began work with the collection in the early 1990s, the Kenrick family had just agreed to house the material at the Black Country Museum. Unfortunately, at that time, none of the collection was cataloged so we cannot offer more specific references. See also *Midland Chronicle and Free Press*, 4 July 1891; and Church, *Kenricks in Hardware*, 25-28.

29. W. E. Jephcott, *The House of Izon: The History of a Pioneer Firm of Ironfounders* (London: Murray-Watson, 1948), 11.

30. William Hawkes Smith, *Birmingham and Its Vicinity as a Manufacturing District* (London: Charles Tilt, 1836), Part III, 31, italics in the original.

31. Smith, *Birmingham and Its Vicinity*, Part III, 30-34.

32. William Kenrick, "The Hollow-Ware Trade," in *Birmingham and the Midland Hardware District*, eds. British Association for the Advancement of Science and Samuel Timmins (London: Cass, 1967 [1866]), 104-105. See also the very detailed description of the enamelling process entitled "Notes on Enamelling at Spon Lane from 1841 to 1893 by Sir George Kenrick," mimeo, n.d., Kenrick Collection.

33. During the first half of the nineteenth century, the distinction between the British "shop" and the "factory" was unclear. Small, detached shops engaged in external subcontracts or "outwork" for larger firms while hundreds of people might work in a large plant without steam power, collected in small work groups.

Children's Employment Commission (1862) [CEC], *Third Report of the Commissioners*, No. 3414-1 (1864), 63. See also Allen, *The Industrial Development of Birmingham*; Sidney Pollard and Paul Robertson, *The British Shipbuilding Industry, 1870-1914* (Cambridge, Mass.: Harvard University Press, 1979).

34. "Stock: September 30th 1829 and June 30th 1830," and "Hiring Book, 1835," Kenrick Collection.

35. Later, the apprentices would sign an "indenture statement," indicating a more formal system of apprenticeship and training. The earliest evidence of this system appears to be 1884 as described in a letter from W. B. Kenrick to Miss I. Humphreys, 8 May 1951, Kenrick Collection.

36. D. C. Woods, "The Operation of the Master and Servants Act in the Black Country, 1858-1875," *Midland History* 7 (1982): 109. While we found no evidence in the firms records or local papers regarding prosecutions at Kenricks, the prevalence and public nature of cases in the area must have provided a clear deterrent to breaking one's contract. See Woods, "The Operation of the Master and Servants Act," for a detailed discussion of the role of the Act in shaping the social relations of production in the Black Country.

37. Although this was certainly not the case across the metal trades. The hiring of girls and young women was concentrated in the button, pin, pen, jewelry, and assorted trades. See Allen, *The Industrial Development of Birmingham* and chapter 4, this volume.

38. CEC, 52-53

39. CEC, 17-18.

40. CEC, testimony of "Messrs. A. Kenrick and Sons', Ironfounders, West Bromwich," 144-145. Text of the complete testimony included in appendix A.

41. Evidence indicates that there were a few "managing" craftsmen in some of the shops by the mid-nineteenth century. In a letter to the directors on the occasion of the firm's centennial celebration, Sarah Sutton reminisces about her father, William Neale, "foremen of the moulders," as well as William Sutton, the "managing turner." Contracts for these men appear in the 1835 Hiring Book, and the conditions of their employment were piece rates. Later, William Neale's name appears as witness to other contracts. It is not clear exactly what the role of these men were or if their compensation differed from the other contractors. That is, we cannot be certain that they were "company" foremen, paid wages by the masters, or had supervisory and disciplinary powers. See letter from Sarah Sutton to John Arthur Kenrick, 23 June 1891, Kenrick Collection.

42. By best estimates, the Kenrick engine would have been one of only a few such engines in operation in the district in 1812. See Court, *Midland Industries*, 252-259. Kenrick valued his engine that year at £340 (Stock Book, 1814, Kenrick Collection).

43. Smith, *Birmingham and Its Vicinity*, 16. Italics in the original.

44. As Berg put it, "A working man could continue to use the same basic tools developed in the eighteenth century and rent room and power to pursue his

trade at greater speed and efficiency." Yet, she does note that, "steam power imposed a much greater regularity on the working day, and even the self-employed artisan could no longer organize his working day around his other familial, cultural and community commitments." Berg, *The Age of Manufactures: Industry, Innovation and Work in Britain, 1790-1820* (New York: Routledge, 1994), 279.

45. Joseph Hood, Articles of Agreement, 30 November 1840, Kenrick Collection.

46. Kenrick, "The Hollow-Ware Trade," 106.

47. Church, *Kenricks in Hardware*, 65.

48. Allen, *The Industrial Development of Birmingham*, 20-23.

49. Church, *Kenricks in Hardware*, 66.

50. CEC, evidence of Messrs. A. Kenrick and Sons', 144. See appendix A.

51. Marx, *Capital* 1: 489.

52. Hartmann, "Capitalism, Patriarchy, and Job Segregation by Sex," 150.

53. Berg, "Women's Work, Mechanization, and the Early Phases of Industrialization in England," 84.

54. Burawoy, "Karl Marx and the Satanic Mills," 251.

55. Littler, *The Development of the Labour Process in Capitalist Societies*, 78 and Eric J. Hobsbawm, *Labouring Men: Studies in the History of Labour* (New York: Basic Books, 1964), 298.

Chapter 3

1. See, for example, Friedrich Engels, *The Condition of the Working Class in England* (Stanford, Calif.: Stanford University Press, 1968 [1844]), and Eric Hobsbawm, *Labouring Men*.

2. W. C. Aitken, "Brass and Brass Manufacturers," in *Birmingham and the Midland Hardware District*, eds. British Association for the Advancement of Science and Samuel Timmins (London: Cass, 1967 [1866]), 363-364.

3. Aitken, "Brass and Brass Manufacturers," 365.

4. *Reports of the Inspectors of Factories to Her Majesty's Principal Secretary of State for the Home Department for the Half Year Ending 31st October 1868*, no. 4093-I (1869), 271.

5. Children's Employment Commission (1862) [CEC], *Third Report of the Commissioners*, no. 3414-1 (1864), report of J. E. White, 144.

6. William Kenrick, "The Hollow-Ware Trade," in *Birmingham, and the Midland Hardware District*, eds. British Association for the Advancement of Science and Samuel Timmins (London: Cass, 1967 [1866]), 108-109. A "freeholder," in contrast to a mere leaseholder, or "copyholder," had the right to pass on his land, or "estate," through inheritance.

7. British Association for the Advancement of Science and Samuel Timmins, eds., *Birmingham and the Midland Hardware District* (London: Cass, 1967 [1866]), 642-643.

8. George J. Barnsby, *Social Conditions in the Black Country, 1800-1900* (Wolverhampton, England: Integrated Publishing Services, 1980), 1-23. Barnsby's book, while admittedly partisan, is very thorough and references many primary sources.

9. Barnsby, *Social Conditions in the Black Country*, 151.

10. *Report to the Local Government Board on the Sanitary Condition of the Urban Sanitary District of West Bromwich* (London: Her Majesty's Stationery Office, 1875).

11. Barnsby, *Social Conditions in the Black Country*, 229.

12. Barnsby, *Social Conditions in the Black Country*, 226-229.

13. Barnsby, *Social Conditions in the Black Country*, 235.

14. CEC, 145.

15. CEC, 145-146.

16. CEC, 146-148.

17. CEC, 53.

18. CEC, 144.

19. Allen, *The Industrial Development of Birmingham*, 199.

20. Kenrick, "The Hollow-Ware Trade," 108.

21. Factory Acts Extension Act and the Workshops Act CAP CXLVI (1867).

22. Factory and Workshops Acts Commission, *Report of the Commissioners*

Appointed To Inquire into the Working of the Factory and Workshops Acts: With a View to Their Consolidation and Amendment: Together with the Minutes of Evidence, Appendix, and Index, vol. 2, *Minutes of Evidence,* C 1443-I (1876), 333. "John Arthur Kenrick Esq., examined," 331-334. Text of the complete testimony included in Appendix B.

23. Factory and Workshops Acts Commission, *Report of the Commissioners,* 332.

24. Douglas Reid, "The Decline of Saint Monday, 1766-1876," *Past and Present* 71 (May 1976): 87.

25. Reid, "The Decline of Saint Monday," 89.

26. CEC, 144. See appendix A.

27. Factory and Workshops Acts Commission. *Report of the Commissioners,* 332.

28. See *Articles of Association,* Kenrick Collection. Seven members were required by the Companies Act; thus, one share was given to Thomas Martineau, son-in-law to Timothy, and the firms solicitor, Herbert Chamberlain, who was William's brother-in-law, and George Kenrick, the son of Archibald II, a farmer in Nottingham. Church, *Kenricks in Hardware,* 78.

29. *The Free Press,* 4 July 1891. Many of Archibald's "recipes" appear in the inside covers of record books, etc. A letter by the daughter of one worker, William Neale, who had been a bearer at Archibald Kenrick's funeral, recalled how the elder Kenrick had visited her father, a moulder, during an illness, and she stated that "Your grandfather was noted as a man of great perseverance in business, but never lost an opportunity of doing a good turn to his work people." Letter from Sarah Sutton to John Arthur Kenrick, 23 June 1891, Kenrick Collection.

30. CEC, 144-145. See appendix A.

31. Joyce, *Work, Society, and Politics,* 137.

32. Pamphlet, Temperance and Educational Institute, n.d., Kenrick Collection.

33. "Indenture of Apprenticeship," Kenrick Collection.

34. *The Midland Chronicle and Free Press,* 10 December 1943; 28 January 1944. See also copies of the *Annual Report of the Birmingham Hospital Saturday Fund* (Birmingham Public Library) 1883 and after. The company discontinued subscriptions in 1897. Of course, much of this benevolence was more symbolic than substantial because the contributions were minimal and all these "benefits" required the financial support of workers. As Church noted, "Thus, with minimal assistance from the employers, the employees were entirely responsible for coping with sickness." See Church, *Kenricks in Hardware,* 279-280. Further, until required by law, compensation for injuries were apparently handled on a case-by-case basis. We discovered evidence of one such instance in which Henry Harris was given the sum of £10 as a "gratuity" in consideration of ". . . the loss sustained by me in the death of my son Abraham Harris who was killed through an accident of his own causing at their works and I declare that I fully absolve Messrs. A.

Kenrick and Sons from all responsibility or liability whatsoever in the matter." Kenrick Collection.

35. *Midland Chronicle and Free Press*, 4 July 1891.

36. See David S. Landes, *Unbound Prometheus: Technological Change and Industrial Development in Western Europe from 1750 to the Present* (Cambridge: Cambridge University Press, 1969), 233-234.

37. Church, *Kenricks in Hardware*, 141-142.

38. Directors' Minutes (hereafter, DM), 30 June 1888. Kenrick Collection; *Midland Chronicle and Free Press*, 4 July 1891.

39. Handscript statement of demands and response, Kenrick Collection.

40. *West Bromwich Weekly News*, 7 February 1890. For a historical account of the Knights' activities see, Henry M. Pelling, "The Knights of Labour in Britain," *Economic History Review* 9, no. 2 (December 1956): 313-331.

41. *The Birmingham Daily Gazette*, 7 February 1890.

42. *The Birmingham Daily Gazette*, 7 February 1890. The men asserted later that J. A. Kenrick actually suggested that the men who would lose "privileges" be compensated knowing full well that his men would not gain such pay.

43. *The Birmingham Daily Gazette*, 7 February 1890.

44. Details of the strike are found in the *West Bromwich Weekly News*, and *The Birmingham Daily Gazette* for the months January through March.

45. *The Birmingham Daily Gazette*, 5 February 1890; 15 February 1890.

46. *The Birmingham Daily Gazette*, 12 March 1890.

47. Agreement with George Brown, 12 March 1890, Kenrick Collection. We found half a dozen of the contracts with these men in the firm's records. By consulting the census records for 1891, we found that the six lived in the immediate area, and with the exception of one, were in their early twenties and were in families of five or more siblings. See *Enumerators' Schedule of Households in West Bromwich*, 1891, Local Studies Centre, Smethwick Library, High Street, Smethwick.

48. DM, 19 March 1890.

49. DM, November-December 1890.

50. Our narrative of the Centennial celebration is derived from the eyewitness coverage of a reporter for the *Midland Chronicle and Free Press*, 4 July 1891. With the exception of the text of the Centennial Address, quotations are from the newspaper; we assume that the reporter was paraphrasing the speakers. The Centennial celebration cost the firm just over £1000 according to DM, 19 August 1891.

51. Church, *Kenricks in Hardware*, 293.

52. DM, 19 August 1891.

53. DM, 16 April 1890.

54. DM, 20 January 1892.

55. We found eight such addendum agreements in the Kenrick Collection.

56. Church, *Kenricks in Hardware*, 292.

57. Letter from Geo. Holloway to Messrs. Archibald Kenrick and Sons, Ltd., 6 October 1896, Kenrick Collection.

58. Letter from W. H. Bennett to Messrs. Archibald Kenrick and Sons, Ltd., October 1896, Kenrick Collection

59. DM, 17 August 1892.

60. Littler, *The Development of the Labour Process in Capitalist Societies*, 79.

61. See P. Cressey and J. MacInnes, "Voting for Ford: Industrial Democracy and the Control of Labour," *Capital and Class* 11 (summer 1980): 5-33, as well as Littler, 79.

62. Church, *Kenricks in Hardware*, 58.

63. Joyce, *Work, Society, and Politics*, 136.

64. Reid, "The Decline of Saint Monday," 99.

65. *The Birmingham Daily Gazette*, 7 February 1890. The men asserted later that J. A. Kenrick actually suggested that the men who would lose "privileges" be compensated knowing full well that his men would not gain such pay.

66. Church, *Kenricks in Hardware*, 292.

67. Joyce, *Work, Society, and Politics*, 68.

68. *The Birmingham Daily Gazette*, 5 February 1890. See also worker White in his Centennial speech when he refers to an "accord that has always existed even when disputes between capital and labour had been general—(applause)."

69. Factory and Workshops Acts Commission. *Report of the Commissioners*, 331; DM, 19 December 1894.

70. Church, *Kenricks in Hardware*, 296. Of course, some skilled occupations were created by the mechanization (e.g., machine tool setters, decete, etc.) but the net change was clearly an increase in the semi-skilled. This trend occurred throughout other industries as well. For example, as Allen notes, the percentage of semiskilled female labor in the tin-plate industry of the Midlands had doubled from 18 percent in 1861 to 37 percent by 1911. See Allen, *The Industrial Development of Birmingham*, 342, and chapter 4, this volume.

71. Joyce, *Work, Society, and Politics*, 65, and Burawoy, "Karl Marx and the Satanic Mills," 260-263.

Chapter 4

1. Hugh Kenrick, "Extracts from Minute Books: Meetings of Directors and Annual Reports Since the Incorporation of the Company in 1883," mimeo, n.d., Kenrick Collection. These "Extracts" are undated, but were compiled near or at the end of his tenure with the firm some time after World War II. While Hugh's intent in compiling this chronological summary was clearly to provide an objective finding aid to the five volumes of Directors' Minutes (DM) (usually an "extract," is a direct quote, or faithful summary, from the DM), as in his description of the purpose of Ryland's report above, he frequently hits the sociological nail square on the head.

2. Kenrick, "Extracts," Kenrick Collection. The 1890s also saw the firm reach its highest levels of sales, profits, and number of workers employed. And between 1883 and 1914 over £600,000 was distributed to Kenrick shareholders in dividends and bonuses, Church, *Kenricks in Hardware*, 143.

3. Barbara Drake, *Women in the Engineering Trades*, London: Fabian Research Department, 1917, and her *Women in Trade Unions*, London: Virago Press, 1984 [1920]; Gail Braybon, *Women Workers in the First World War: The British Experience* (London: Croom Helm, 1981); John, ed. *Unequal Opportunities*; Rose, "Gender Antagonism" and *Limited Livelihoods*, 30-33.

4. Edward Cadbury, M. Cecile Matheson, and George Shann, *Women's Work and Wages: A Phase of Life in an Industrial City* (Chicago: University of Chicago Press, 1907), 132-133.

5. The Kenricks were relative latecomers to the practice of deskilling, mechanizing, and hiring unskilled women to replace skilled men. In her useful discussion of women workers in the small metal trades in Birmingham and the Black Country between 1750 and 1850, Ivy Pinchbeck writes, "The introduction of machinery in these trades in the early nineteenth century was accompanied by an increase in the proportion of women employed in them . . . Machinery and a series of trade depressions which permitted the exploitation of women's labor appear to have been responsible for this increase," Ivy Pinchbeck, *Women Workers and the Industrial Revolution, 1750-1850* (London: Cass, 1969 [1930]), 275. And Maxine Berg writes, referring to the late eighteenth century, "It is said, however, that the adoption of machines for stamping and piercing extended the range of female employment especially for young girls. And it was recognized that women's work was widespread in the japanning and the stamping and piercing trades. Girls were specifically requested in advertisements for button-piercers, annealers, and stoving and polishing work in the japanning trades." See Berg, "Women's Work, Mechanization, and the Early Phases of Industrialization in England," 85. The turning of cast-iron hollow-ware was apparently somewhat less amenable to deskilling and mechanization than were the less complicated tasks of making buttons, nails, washers, screws, hooks and eyes, and so forth to which Pinchbeck was referring, al-

though the Kenricks did consider mechanization as early as the 1840s but decided against it at that time for a variety of reasons (see chapter 2, this volume).

6.   Kenrick, "The Hollow-Ware Trade," 108.

7.   Reading through the DM for the years between 1892 and 1913 one finds the directors voting thousands of pounds to be used for "tearing down," "building," "converting," "buying," and "replacing" plant and machinery.

8.   The gradual substitution of enamelled steel, and later aluminum, brought about the decline of the cast-iron hollow-ware industry during the last decade of the nineteenth century. Kenricks responded by restricting trade via the Cast Iron Hollow-Ware Makers Association, by absorbing small competitors out to undersell them, and by diversifying their line of products—often by buying out small firms in closely related metal trades. In 1887, Kenricks acquired a controlling interest in A. & E. Baldwin and Sons Ltd., which included their hinge-making subsidiary, the Anglo American Tin Stamping Co.; they bought out the Birmingham Hollow-Ware Company in 1887; the Glasgow firm of W. Nielson in 1888; and J. B. Bond in 1892. Between 1900 and 1910 Kenricks expanded the hinge-making part of their business by acquiring the Lionel Street Company, the E. K. and G. A. Martineau & Co., and the Dial Hinge Company. All these hinge-making enterprises were, in 1910, organized as United Hinges Ltd. and located adjacent to the Kenrick works in West Bromwich. See Church, *Kenricks in Hardware*, 125-133.

9.   As will be discussed below, the Kenricks' main factory and its subsidiaries made a variety of metal products for the Ministry of Munitions of War during World War I, including shells, grenades, and mess tins. As such, they were a "controlled establishment" and subject to considerable government monitoring and regulation via the Ministry. One thing they were apparently required to do was periodically fill out questionnaires on their workforce, several copies of which are contained in the Kenrick Collection. Thus, the figure of 33 percent comes from a questionnaire, dated 7 November 1918, that asked the employers to provide data on the workforce on the eve of war, and also for July of 1918, by which time the percentage of women was reported to have risen to 59 percent (549 of 929). Another questionnaire, dated 1 March 1916, shows the percentage of women to be 41 percent (373 of 904).

10. The deskilling also brought an end to male apprenticeship at Kenricks by 1903, when the last apprentice was taken on. In a letter to Miss I. Humphreys, on 8 May 1951, then Chairman William B. Kenrick wrote, "The Cessation of Apprenticeship in Hollow-ware Turning was influenced by the substitution of a Mechanical devise for the Skilled Handicraft and both in Turning and Tinning by the decay of the Cast Iron Hollow-ware trade due to its gradual substitution by Hollow-ware made of Enameled Steel or Aluminum." Letter from William B. Kenrick to Miss I. Humphreys, 8 May 1951, Kenrick Collection.

11. Braybon, *Women Workers*, 24-32. See also Sheila Lewenhak, *Women and Trade Unions: An Outline History of Women in the British Trade Union Move-*

ment (New York: St. Martin's Press, 1977); Norbert C. Soldon, *Women in British Trade Unions, 1874-1976* (Totowa, N.J.: Rowman & Littlefield, 1978); Sarah Boston, *Women Workers and the Trade Union Movement* (London: Davis-Poynter, 1980); E. H. Hunt, *British Labour History, 1815-1914* (Atlantic Highlands, N.J.: Humanities Press, 1981). Referring to the 1890-1910 period, Drake wrote, "Trade unionism was most backward amongst women metal and wood-workers. The problem of female labour was so far not an acute one except in the Birmingham and Black Country small metal trades." Drake, *Women in Trade Unions*, 38.

12. See Hartmann, "Capitalism, Patriarchy, and Job Segregation by Sex"; Rose, "Gender Antagonism."

13. Louise Tilly and Joan Wallach Scott, *Women, Work, and Family* (New York: Holt, Rinehart and Winston, 1978); Hunt, *British Labour History*, 25.

14. Drake, *Women in Trade Unions*, 31; Sally Alexander, *Women's Work in Nineteenth-Century London: A Study of the Years 1820-50* (London: Journeyman Press and London History Workshop Centre, 1983); Rose, "Gender at Work," 114-117 and *Limited Livelihoods*, 22-49; Elizabeth Roberts and Economic History Society, *Women's Work 1840-1940* (Basingstoke: Macmillan, 1988), 13; Lewenhak, *Women and Trade Unions*, 29-44; Hunt, *British Labour History*, 17-25; Standish Meacham, *A Life Apart: The English Working Class, 1890-1914* (Cambridge, Mass.: Harvard University Press, 1977), 103 .

15. Rose, "Gender Antagonism," 195.

16. Meacham, *A Life Apart*, 103. For a comparative study of factory politics in engineering in the United States and Britain between 1890 and 1922 that discusses gender, see Haydu, *Between Craft and Class*.

17. Meacham, *A Life Apart*, 103, 60-115.

18. Cadbury, Matheson, and Shann, *Women's Work and Wages*, 39-40. Some of these trades, such as enamelling, tin-plate pressing and stamping, and brass casting, were employed at Kenricks.

19. Braybon, *Women Workers*, 30.

20. Braybon, *Women Workers*, 28; see also Drake, *Women in the Engineering Trades*, 7-8, and Dennis Smith, "Paternalism, Craft and Organizational Rationality 1830-1930: An Exploratory Model," *Urban History* 19, pt. 2 (October 1992): 211-228.

21. Memorandum from J. E. Harstow and H. C. Everett to Chief Inspector regarding Welfare Order: Hollow-Ware, 26 March 1919, in file "Hollow-Ware and Galvanized Welfare Order, 1921," Ministry of Labor PRO/LAB/14/203.

22. Church, *Kenricks in Hardware*, 277.

23. Marx, *Capital*, 1: 291.

24. "The Midland Employers' Federation Report of Special General Meeting, March 30th, 1916." MRC/MSS.265/M/3/2. A "Mr. Kenrick," probably either Sir George Kenrick, the firm's chairman during this time, or Clive Kenrick, works manager in charge of labor relations, was in attendance.

25. Barnsby, *Social Conditions*, 229.

26. Barnsby, *Social Conditions*, 229.

27. Barnsby, *Social Conditions*, 229. See also Richard H. Trainor, *Black Country Elites: The Exercise of Authority in an Industrialized Area, 1830-1900* (Oxford: Clarendon Press, 1993), 52-53.

28. Wally Seccombe, "Patriarchy Stabilized: The Construction of the Male Breadwinner Wage Norm in Nineteenth-Century Britain," *Social History* 2, no.1 (January 1986): 73. As Cadbury, Matheson, and Shann wrote, "The artisan's wife is very thankful when the boys and girls begin to earn, but she does not wait for this in order to feed them properly." Cadbury, Matheson, and Shann, *Women's Work and Wages*, 235.

29. The population of West Bromwich in 1901 was about 65,000 of which most—at least 50,000—were dependent upon wage labor. Since average household size at the time was about 5, there were approximately 10,000 working class households in West Bromwich at the turn of the century. See Trainor, *Black Country Elites*, 47 and Barnsby, *Social Conditions*, 84.

30. Trainor notes that by 1900, despite the vigorous prosecution of the parents of truants in West Bromwich, a minority of working class families, and no doubt the poorest among them nevertheless ". . . flouted the compulsory attendance clauses of the education acts," Trainor, *Black Country Elites*, 171, 275.

31. Eric Hobsbawm suggests that the declining prospects of such young men as these led to labor militance in 1910-1914. He states, "On the other hand the fairly extensive de-skilling which took place in the last 30 years before 1914 created the frustration which Askwith, the government's chief industrial conciliator in those years, thought important. The young worker: '. . . does not like to admit to himself that he is not being trained as an engineer or a shipbuilder or a housebuilder, but to become an operative. But in a brief time to the majority comes disillusionment; and then once a man is disillusioned, bitterness is a very natural result, and antagonism to the system which he deems to be the cause.'" See Eric J. Hobsbawm, *Workers: Worlds of Labor* (New York: Pantheon, 1984), 206.

32. Cadbury, Matheson, and Shann, *Women's Work and Wages*, 46.

33. Cadbury, Matheson, and Shann, *Women's Work and Wages*, 52.

34. Cadbury, Matheson, and Shann, *Women's Work and Wages*, 50-118.

35. From E. Sylvia Pankhurst, *The Suffragette*, as quoted in E. H. Phelps Brown, *The Growth of British Industrial Relations: A Study from the Standpoint of 1906-1914* (London: Macmillan, 1965), 69.

36. Cadbury, Matheson, and Shann, *Women's Work and Wages*, 113-118.

37. Cadbury, Matheson, and Shann write, "There is a very general complaint that girls will not learn a trade because, in the first place, they all hope to marry and henceforth to be under no necessity of earning their own living. One would think that the existence of such a vast number of married women working in factories in a city like Birmingham would have a sobering effect on such speculations," Cadbury, Matheson, and Shann, *Women's Work and Wages*, 46.

38. Cadbury, Matheson, and Shann, *Women's Work and Wages*, 115-116.

39. Barnsby, *Social Conditions*, 84-99.

40. Cadbury, Matheson, and Shann, *Women's Work and Wages*, 200.

41. Marx, *Capital*, 1: 449. See Harold Benenson, "The Family Wage and Working Women's Consciousness in Britain, 1880-1914," *Politics and Society* 19, no. 1 (1991): 93; and Cadbury, Matheson, and Shann, *Women's Work and Wages*.

42. "The girls are naturally averse to revealing their private affairs, and often even a girl's mother has no definite knowledge of the exact wages earned by the girl. The girls were also nervous lest through giving information they should get into trouble with their employers. They were often suspicious as to the motives of their inquirer. After a most careful explanation of the aim of the investigation one would hear such remarks as, "What cheek to ask us what we earn." "They are going to take our answers to the masters to get our wages lowered." Cadbury, Matheson, and Shann, *Women's Work and Wages*, 14.

43. Cadbury, Matheson, and Shann, *Women's Work and Wages*, 238-239.

44. Church, *Kenricks in Hardware*, 84.

45. Littler, *The Development of the Labour Process in Capitalist Societies*, 64-79.

46. Littler, *The Development of the Labour Process in Capitalist Societies*, 81. Church reports evidence that suggests the Kenricks were careful to inject petty capitalist motivations down through the managerial hierarchy, "In the absence of detailed wage figures specifying remuneration by grade of employee, it is not possible to comment on the particular relationships between foreman and piece-workers. A record remains of a foreman, Barrett who, in addition to his own wage, was to receive 17½ per cent of the total wages paid to nine enamellers, an arrangement which suggests that in one department, at least, foremen's earnings were geared directly to output but not at the pieceworker's expense." See Church, *Kenricks in Hardware*, 84.

47. Church, *Kenricks in Hardware*, 84. We have only an exact figure for 1918, which shows 28 male foremen. But by this time the firm's labor force had shrunk from its high in the late 1890s of about 1,250 to about 900 in 1920. Thus, given a constant proportion of foremen to workforce, there should have been approximately 40 foremen near the turn of the century. See also Church, *Kenricks in Hardware*, 143, note 1.

48. Littler, *The Development of the Labour Process in Capitalist Societies*, 86. We have no evidence of any attempt by the directors to sweep away their traditional approach and replace it with anything "scientific." Indeed, in the 1930s a consultant's report will criticize the firm for its failure to more fully commit itself to workplace reorganization, rationalization, and mass production. See Church, *Kenricks in Hardware*, 201-218.

49. Littler, *The Development of the Labour Process in Capitalist Societies*, 86.

50. James Leask and Philomena Bellars, *"Nor Shall the Sword Sleep . . ."*: *An Account of Industrial Struggle*, Birmingham Public Library: n.p., 1954.

51. Not surprisingly, the Kenricks' version of the events, published a few days later, presents a more reasonable foreman, but they too acknowledge his authority to make decisions about who would do what work, and how they would be paid for it: "On Monday morning one of the helpers in the rolling department at United Hinges Ltd., who had been working days, and who was asked to go on the night turn, objected. He saw his foreman during his breakfast half-hour. The foreman discussed the matter for some time. It being breakfast time, the man suggested that he should come again after he had his breakfast, at nine o'clock. The helpers then left the works, and instead of returning as proposed, persuaded all the other workers in their department to leave also." See *The Free Press*, 18 April 1913.

52. Littler, *The Development of the Labour Process in Capitalist Societies*, 83-84. Littler describes the Premium Bonus System (PBS) as follows: "All PBSs have common elements. Essentially they substitute a time allowance for a piece-work price. This entails fixing time limits for each and every job . . . A minimum rate was guaranteed, and if a job was accomplished in less than the time limit, a worker was paid over and above his standard wage for any time saved, but the bonus was always proportionate. Thus, the 'time saved' was divided between workers and management in varying proportions . . . such that the worker's share of the time saved was a constantly diminishing proportion. Indeed, beyond a certain mid-point bonus earned actually declines . . . Clearly, this was one of its advantages to the employer, because it institutionalized rate-cutting."

53. "Rules to Be Observed in the Works of Messrs. Archibald Kenrick & Sons." Handwritten, no author, no date. T & C Clark Collection, DB/4/11, at the Wolverhampton Borough Archives and Local Studies Centre, Wolverhampton. The hours, meal times, and days of work correspond to the description provided by Arthur Wharton when he first joined the firm in 1899. By 1902, workers at Baldwins, a Kenrick subsidiary, petitioned and won a reduction in hours from 57 to 53 (DM, 19 March 1902), as did the workers at the main factory on Spon Lane after a strike in 1913. Hours were reduced to 47 after World War I, where they would stay until the mid-1920s, when too few, rather than too many, hours became the problem.

54. Such disciplinary systems as we find at Kenricks were apparently widely used in the metal trades. See Cadbury, Matheson, and Shann, *Women's Work and Wages*, 200-209.

55. Church, *Kenricks in Hardware*, 222-223, 316, note 4.

56. R. A. Church, "Family and Failure: Archibald Kenrick and Sons Ltd., 1900-1950," in *Essays in British Business History*, ed. Barry Supple (Oxford: Clarendon Press, 1977), 113.

57. Church, *Kenricks in Hardware*, 280-283.

58. Where possible, foremen would be sent "... back to the bench." DM, 10 August 1921.

59. Littler, *The Development of the Labour Process in Capitalist Societies*, 90.

60. With respect to the nurses, Church writes, "Thus when in 1891 Frederick Ryland recommended the appointment of a trained nurse to visit cases of sickness among the company's employees, the board agreed to offer a post for one year at a salary not exceeding £80 'in the hope that at the end of that time the workpeople will join the company in defraying the expense.' In fact, two nurses accepted a joint appointment at £80 per annum, but their continued employment remains in doubt." Church, *Kenricks in Hardware*, 282

61. Church, *Kenricks in Hardware*, 283.

62. Trainor, *Black Country Elites*, 151.

63. DM, various dates.

64. Other factors such as oligopoly, colonial markets, peak in demand for products, played a part.

65. Church, *Kenricks in Hardware*, 292.

66. Clegg notes that fourteen of the four thousand strikes accounted for fifty-one million of the seventy million working days lost during the period, and did the most to earn the period its rebellious reputation. The coal miners begin the period, and Clegg's list of the top fourteen strikes, with a strike in September of 1910 (13,000 strikers; 2,985,000 working days lost), they conclude it with a strike in February of 1914 (150,000 strikers; 2,654,000 working days lost), and they also contributed the largest strike by far: the national strike of February, 1912 (1,000,000 strikers; 30,800,000 working days lost). The remaining strikes were: seamen and dockers, 1911 (120,000 strikers; 1,020,000 working days lost); dockers and carmen, 1911 (77,000 strikers; 500,000 working days lost); railwaymen, 1911 (145,000 strikers; 485,000 working days lost); dockers and seamen, 1911 (48,000 strikers; 376,000 working days lost); cotton weavers, 1911 (160,000 strikers; 2,954,000 working days lost); jute workers, 1912 (28,000 strikers; 726,000 working days lost); dockers and carters, 1912 (100,000 strikers; 2,700,000 working days lost); cab drivers, 1913 (11,000 strikers; 637,000 working days lost); tube and metal workers, 1913 (i.e., Black Country strike, 50,000 strikers; 1,400,000 working days lost); transport workers, 1913 (20,000 strikers; 1,900,000 working days lost); construction workers, 1914 (20,000 strikers; 2,654,000 working days lost). See Hugh A. Clegg, *A History of British Trade Unions since 1889*, vol. 2, *1911-1933* (Oxford: Clarendon Press, 1985), 24-26.

67. George Dangerfield, *The Strange Death of Liberal England* (New York: Capricorn Books, 1961 [1935]), 217.

68. Arthur J. Taylor, "The Economy," in *Edwardian England 1901-1914*, ed. Simon Nowell-Smith (London: Oxford University Press, 1964), 129-131. On the earlier, more prosperous period, Hobsbawm writes, "Three factors affected the workers' material conditions of life after 1870: the dramatic fall of the cost of

living during the so-called Great Depression of 1873-96, the discovery of the domestic mass market, including that of the well-paid or at least regularly employed workers for industrial produced or processed goods, and (after 1875) the so called 'by-law housing' (under section 157 of the Public Health Act), which in fact created so much of the environment of working-class life, the rows of terraced houses outside the old town centres. All imply or were based on, the modest, patchy, but plainly undeniable improvement in the standard of life of the bulk of British workers, which is not a matter of dispute even among historians." Hobsbawm, *Workers*, 201.

69. Hobsbawm, *Workers*, 210.

70. Taylor, "The Economy," 129-131.

71. Hobsbawm, *Workers*, 207; see also Taylor, "The Economy," 133-134; Dangerfield, *The Strange Death,* 235; and Standish Meacham "The Sense of an Impending Clash: English Working-Class Unrest before the First World War." *American Historical Review* 77, no. 5 (December 1972): 1343-1364.

72. Dangerfield, *The Strange Death,* 233-234.

73. Dangerfield, *The Strange Death,* 221.

74. Church, *Kenricks in Hardware*, 293-295; Leask and Bellars, "*Nor Shall the Sword Sleep*," 8-9.

75. Church, *Kenricks in Hardware*, 294.

76. Clegg, *A History of British Trade Unions*, 58.

77. John Corbett, *The Birmingham Trades Council, 1866-1966* (London: Lawrence and Wishart, 1966), 100; Lewenhak, *Women and Trade Unions*, 141. In an interview with John Whiston, WU organizer, historian Richard Hyman took the following notes regarding WU organizing methods: "Would unite at factory gate at lunchtime, hand out leaflets showing WU benefits and explain what WU stood for. Try to talk to men having lunch outside gates, or coming back from lunch. Would call meeting at night, often in pub. Open-air meetings could be held outside gates, but in early days men often afraid to be seen listening. Attempts made to persuade non-members to join; if they refused to talk, would often follow them home. Two or three times, John Whiston fined for persistently following." Richard Hyman Papers, MRC/MSS.51/3/1/54.

78. It was the WU practice to encourage such "amateur" organizers according to Richard Hyman, *The Workers' Union* (Oxford: Clarendon Press, 1971), 76.

79. Hyman, *The Workers' Union*, 51. Since Tangye's was a member of the Birmingham EEF, this strike was the focus of considerable interest by this engineering employers' association, who at an emergency committee meeting to deal with strike instructed the firm to raise wages but refused to concede the principle of a minimum. "Minutes, February 24th, 1913." Birmingham and District Engineering Trade Employers' Association (hereafter BDEA), MRC/MSS.265/B/1/2.

80. *Birmingham Gazette*, 10 April 1913.

81. *The Free Press*, 18 April 1913.

82. *Birmingham Gazette*, 8 April 1913.

83. Leask and Bellars, *"Nor Shall the Sword Sleep,"* 9. P. Loxston, a contemporary, sized up Geobey and Varley this way: "George Geobey, a district organizer, was a real left, *left* wing light and very good for soap box material and factory gate meetings. Never afraid to tackle any task, pleasant or unpleasant. Julia Varley, a nationally known figure, was a tower of strength on behalf of the female members. She was most capable, either in the board room or amongst the rank and file." Letter from P. Loxston to Richard Hyman, 1 December 1965, Richard Hyman Papers MRC/MSS.51/3/1/36. Very little has been written about Geobey, but Varley has received some attention from historians. Thom's comparison of Mary MacArthur and Julia Varley provides the best available discussion of Varley and her work for the Workers' Union. The Julia Varley Papers are located at the University of Hull. See Deborah Thom, "The Bundle of Sticks: Women, Trade Unionists and Collective Organization before 1918," in *Unequal Opportunities: Women's Employment in England 1800-1918*, ed. Angela V. John (Oxford: Basil Blackwell, 1985), 261-289. See also in the same volume, Ellen Mappen, "Strategists for Change: Social Feminist Approaches to the Problems of Women's Work," 236-259.

84. Details of the strike can be found in the *Birmingham Gazette* and *The Free Press.*

85. Leask and Bellars, *"Nor Shall the Sword Sleep,"* 10.

86. Leask and Bellars, *"Nor Shall the Sword Sleep,"* 9.

87. DM, 30 June 1913. In addition, in his letter to Miss I. Humphreys, on 8 May 1951, William B. Kenrick offered this retrospective assessment of the firm's performance since incorporation: "In general terms the results of the trade may be summarized as follows: 1871-1873 Bad Years; 1875-1879 Good Years; 1881-1883 Bad Years; 1885-1892 Bad or Poor Years; 1893-1903 Improving to Good; 1904-1911 Good; 1912-1914 Bad. In the period of Bad years the worst years were: 1873, 1883, 1890, and 1914." Letter from William B. Kenrick to Miss I. Humphreys, 8 May 1951, Kenrick Collection.

88. Church, *Kenricks in Hardware*, 294.

89. Leask and Bellars, *"Nor Shall the Sword Sleep,"* 7.

90. We quote from the account at length because it is as close to the views of the strikers as we have been able to get. Virtually everything else we have involves someone else speaking to them, for them, or about them, which makes it very difficult to separate the workers' expression of their interests from the constructions of their interests by others.

91. The demand for the minimum also implied a demand that the men in the cold-rolling mill be paid 23s whether on the day or night shift.

92. While these workers were employed at Guest, Keen, & Nettlefold, rather than Kenricks, management practices and working conditions at the two factories probably differed very little. John Arthur Kenrick (1829-1926), who was chairman of the Kenrick Board of Directors from1883 to1903, was also appointed to the Nettlefolds Ltd. Board of Directors, serving as its chairman at the time it fused

with Guest and Keen in 1902. In addition, John Arthur's son, John Archibald Kenrick (1869-1933), subsequently married Grace Nettlefold, daughter of E. J. Nettlefold, a Nettlefold's senior partner. See Edgar Jones, *A History of GKN*, vol. 1, *Innovation and Enterprise, 1759-1918* (London: Macmillan, 1987). And as noted previously, John Archibald was the Kenrick director directly responsible for labor relations during the 1913 strike.

93. Hyman, *The Workers' Union*, 52.

94. Meacham, *A Life Apart*, 60-115; Braybon, *Women Workers*, 24. This was a continuation of the debate over the "women question," that had been going on at least since 1800, perhaps given additional urgency by the spread and greater visibility of "factory girls" beyond the textile and clothing industry.

95. A story about Varley in the Workers' Union *Record* reported, "Believing that where women work with men they should be organized in the same union, she finds the Workers' Union more fully in accord with those ideas, and when her work with the voluntary committee came to a close she had little difficulty in accepting the position our union had to offer." Workers' Union *Record*, 7 July 1914.

96. Corbett, *The Birmingham Trades Council*, 91, 98.

97. Leask and Bellars, *"Nor Shall the Sword Sleep,"* 8. The "Union enquiry" into the conditions of the female workers at Kenricks was likely part of, or the same as, the investigation carried out by Varley and the Birmingham Trades Council given Varley's prominence in both organizations and also that the WU had been a member of the Trades Council since 1905. Corbett, *The Birmingham Trades Council*, 91.

98. Hyman, *The Workers' Union,* 52.

99. Hyman, *The Workers' Union,* 52.

100. *The Midland Chronicle for West Bromwich and Oldbury*, 11 April 1913.

101. Workers' Union *Record*, July 1916, 6-7.

102. Church, *Kenricks in Hardware*, 294; Leask and Bellars, *"Nor Shall the Sword Sleep,"* 8-10; Hyman, *The Workers' Union*, 51-52.

103. Quoted in Leask and Bellars, *"Nor Shall the Sword Sleep,"* 29.

104. In coming years the WU would evolve as a relatively conservative and reformist union (compared to, for example, the coal miners), and Julia Varley would be singled out, among women trade unionists, for her moderate positions and belief in employer-employee cooperation.

105. It is not as if the principle of "equal pay for equal work," as expressed by these workers, was inconceivable in 1913. Far from it. As Drake, *Women in Trade Unions,* 227, wrote: "At the Trades Union Congress of 1888, it was formally resolved for the first time that 'where women do the same work as men they shall receive equal pay,' and mainly with this object in view men have assisted women to organize. The principle of 'equal pay' has since then received practically unanimous support from men and women trade unionists . . . In practice,

however, the policy has been rarely carried out, and men in despair have continued to exclude women so far as possible from employment." And when that proved impossible, as our data show, the strategy was to allow women into the union on the same, unequal, basis upon which employers brought them into the factories.

106. Downs, *Manufacturing Inequality*, 316.

107. Downs, *Manufacturing Inequality*, 316.

108. Quoted in Leask and Bellars, *"Nor Shall the Sword Sleep,"* 24.

109. George Askwith, *Industrial Problems and Disputes* (London: John Murray, 1920), 256.

110. While it is true that the strikers received widespread public support, at least some of the citizenry might have been offended by the spectacle of hundreds of "factory girls" on the march. See Lisa Tickner, *The Spectacle of Women: Imagery of the Suffrage Campaign, 1907-14* (Chicago: University of Chicago Press, 1988). Harrison reports that by 1909 people were referring to Birmingham ". . . as a sort of Mecca for the Anti-Suffragists . . ." and that "Already by 1909 militant conduct had made it difficult for non-militant suffragists as well as suffragettes to move freely round Birmingham." Brian Harrison, *Separate Spheres: The Opposition to Women's Suffrage in Britain* (New York: Holmes and Meier, 1978), 121, 187.

111. Clegg's *A History of British Trade Unions* is only a recent example.

112. On the question of working men benefitting from the labor of working women in the home, suffragette Annie Kennedy observed in 1911, "I saw men, women, boys and girls, all working hard during the day in the same, hot, stifling factories . . . Then when work was over I noticed that it was the mothers who hurried home, who fetched the children that had been put out to nurse, prepared the tea for the husband, did the cleaning, baking, washing, sewing, and nursing. I noticed that when the husband came home, his day's work was over; he took his tea and then went to join his friends in the club or in the public house, or on the cricket or football field, and I used to ask myself why this was so." From E. Sylvia Pankhurst, *The Suffragette*, 22, as quoted in Brown, *The Growth of British Industrial Relations*, 69.

113. Women's employment outside the home, even at a low(er) wage, may have, in the long run, undermined patriarchy at home.

114. Similarly, Benenson argues that ". . . the family wage argument cemented a *common interest* between paternalist employers and male trade unionists . . . integrating men trade unionists into the parliamentary system," Benenson, "The Family Wage," 97, italics in the original. See also Braybon, *Women Workers*, 30.

115. As Barrett and McIntosh put it, "It appears, however, that the organizations of the working class colluded with pressure from the bourgeoisie to structure the working population along the lines of gender," Michelle Barrett and Mary McIntosh, "The 'Family Wage': Some Problems for Socialists and Feminists," *Capital and Class* 11 (summer 1980): 54.

116. Lewenhak, *Women and Trade Unions*, 140. In the process of explaining how support for the "family-wage" arose in nineteenth-century England, Seccombe nicely summarizes the story when he writes, "In response to wage undercutting through the hiring of women workers, most unions pushed for their restriction, curtailment or outright exclusion in the name of a clear priority for the employment rights of men, the great majority of their members. The campaign never succeeded in driving women from industry, but it did manage to seal their marginalization in low wage, high turnover job ghettos," Seccombe, "Patriarchy Stabilized," 73.

117. Sylvia Walby, *Patriarchy at Work: Patriarchal and Capitalist Relations in Employment* (Cambridge: Polity Press, 1986) and *Theorizing Patriarchy* (Oxford: Basil Blackwell, 1990).

118. As the Amalgamated Engineering Union (AEU) did until 1943, according to Noreen Branson. See her introduction to Barbara Drake's, *Women in Trade Unions,* xiii; see also James B. Jefferys, *The Story of the Engineers, 1800-1945* (London: Lawrence and Wishart, 1946).

119. Church, *Kenricks in Hardware*, 143.

120. Askwith, *Industrial Problems and Disputes*, 253. Perhaps they shouldn't have been surprised. Engineering employers, some of whom hired unskilled workers in addition to engineers, could see trouble brewing. In his report for 1912, the secretary for the Birmingham District Engineering Trade Employers' Association wrote, "Some trouble has been occasioned by the action of the Workers' Union in approaching two members with reference to rates of wages paid to men in that society. At the moment no complaints are before the association, but from the activity this society is displaying with firms outside the association it is evident that trouble is looming in the future," "Secretary's Report, 1912," BDEA, MRC/ MSS.265/B/1/2.

121. Lown, "Not So Much a Factory," 34.

122. The Kenricks apparently did not join the MEF immediately, although the issue was under consideration at the 13 August 1913 meeting of the Board of Directors, at which point it was decided to "continue negotiations" on the question of dues and the structure of the management board (DM, 13 August 1913). They joined, probably sooner rather than later, some time between August 1913 and 15 March 1915, because we have evidence showing Clive Kenrick serving on the Management Board as of the latter date. "Executive Committee Meeting, March 15th, 1915," BDEA, MRC/ MSS.265/B/1/3.

123. Hyman, *The Workers' Union*, 56. On the founding of the MEF see also Askwith, *Industrial Problems and Disputes*, 253; Askwith to the Ministry of Reconstruction, "Confidential Memorandum on Employers' Associations," 30 August 1917, PRO/RECO1/376; Leask and Bellars, "*Nor Shall the Sword Sleep*," 19; "The Midland Employers' Federation Report of the Annual General Meeting, January 4th,1917," BDEA, MRC/MSS.265/M/3/2.

124. Hyman, *The Workers' Union*, 56.

125. "Agreement between the Midland Employers' Federation and the Workers' Union, the National Union of Gas Workers, and the Amalgamated Workers, Brickmakers and General Laborers, July 7th, 1913," BDEA, MRC/MSS.265/M/3/1.

Chapter 5

1. Church, *Kenricks in Hardware*, 295.
2. *The Free Press*, 28 November 1913.
3. Church goes on to say, "Thereafter, union support in the district suffered a decline, for the arson case, according to Jones, brought it into local disrepute." Church, *Kenricks in Hardware*, 295.
4. "Minutes, Management Board, March 24th, 1920," Birmingham District Engineering Trade Employers' Association [BDEA]. At the Modern Records Centre [MRC], University of Warwick, MSS.265/B/1/5.
5. Leslie Hannah, *The Rise of the Corporate Economy: The British Experience* (Baltimore: Johns Hopkins University Press, 1976).
6. By 1917, the number of members had grown to 248, but as late as July of 1922, the leaders of the "Allied Trades" section of the Birmingham branch of the Engineering Employers' Federation (which the MEF had joined in 1918), complained that 60 percent of the nonengineering metal trades firms in the district still refused to affiliate with the employers' organization. See "Annual General Meeting, January 4th, 1917," BDEA, MRC/MSS.265/M/3/2; Transcript of meeting between representatives of the Allied Trades employers and Sir Allan Smith, 20 July 1922, "Allied Trades Employers Assoc. Reorganization, Oct., 1921-December, 1923," MRC/MSS.265/B/3/3. For Askwith's description of the MEF, see G. R. Askwith to the Ministry of Reconstruction, "Confidential Memorandum on Employers' Associations," 30 August 1917, PRO/RECO1/376.
7. DM, 13 August 1913.
8. We do not know the exact date they joined, but it must have occurred before 30 March 1916, at which point we find "Mr. Kenrick" of the hollow-ware trade voted onto a committee to investigate the impact of the Munitions Act on the membership. "Special General Meeting, March 30th, 1916," BDEA, MRC/MSS.265/M/3/2.
9. The Engineering Employers' Federation (EEF) was founded in 1896. To simplify the discussion, we will henceforth refer to the national Engineering Employers' Federation as the "EEF" and the Birmingham District Engineering Trades Employers' Association as the "Birmingham EEF," "Birmingham Engineering Employers," or "Birmingham Employers."
10. Later named the Amalgamated Engineering Union (AEU).
11. "Meeting of the Emergency Committee, February 24th, 1913"; "Executive Committee Meeting, September 9th, 1913," BDEA, MRC/MSS.265/B/1/2.
12. "Executive Committee Meeting, August 20th, 1913," BDEA, MRC/MSS.265/B/1/2. The influence of the MEF on the Birmingham engineering employers was also noted in the Secretary's Annual Report, 1912-1913: "You are aware that a new Employers' Association called the Midland Employers' Federation was formed in this District in the early part of the year to deal with questions concerning labor and wages, and that in settlement of certain strikes a schedule of

rates was agreed as applying to youths and Girls. Since this schedule was accepted the Workers' Union made a demand on your Association for the recognition of a similar schedule but with rates slightly higher than those agreed with the Midland Employers' Federation. After discussion in Conference you agreed to observe the same rates as those of the Midland Employers' Federation, and were successful in also including certain provisions in your interests," "Secretary's Report: 1912-13," BDEA, MRC/MSS.265/M/3/1.

13. "Executive Committee Meeting, March 15th, 1915," BDEA, MRC/MSS.265/B/1/2.

14. "Executive Committee Meeting, March 15th, 1915," BDEA, MRC/MSS.265/B/1/2; "Executive Committee Meeting, April 22nd, 1918," BDEA, MRC/MSS.265/B/1/5; and "First Annual General Meeting, February 17th, 1919," BDEA, MRC/MSS.265/B/1/5.

15. We are not suggesting that the Kenricks were class *un*conscious prior to joining the MEF, but we think the firm's willingness to join an organization explicitly established to fight a union marks a turning point in the firm's history. As discussed earlier, the Kenricks belonged to the Cast Iron Hollow-Ware Manufacturers Association (CIHMA) for many years, but this employers' organization was established to regulate the market for finished products rather than, as the MEF, fight with unions about wages. Indeed, as the Rules and Regulations for the CIHMA for 1898-1899 show, any resolutions passed regarding wages were not binding on the membership: "It shall be competent for the Association to discuss the question of wages, but so that no resolution passed on the question shall be absolutely binding on the individual members," "Rules and Regulations of the Association of Cast-Iron Hollow-ware Makers, 1898-99," Kenrick Collection.

16. For the history of politics within the EEF during this period, see Jonathan Zeitlin, "The Labour Strategies of British Engineering Employers, 1890-1922," in *Managerial Strategies and Industrial Relations: An Historical Comparative Study*, eds. Howard F. Gospel and Craig R. Littler (London: Heinemann, 1983), 25-54; "The Triumph of Adversarial Bargaining: Industrial Relations in British Engineering, 1880-1939," *Politics and Society*, 18, no. 3 (1990): 405-426"; and "The Internal Politics of Employer Organization: The Engineering Employers' Federation 1896-1939," in *The Power To Manage? Employers and Industrial Relations in Comparative Perspective*, eds. Steven Tolliday and Jonathan Zeitlin (London: Routledge, 1991), 53-79.

17. See our discussion in the previous chapter on the sources of the industrial unrest between 1910-1914.

18. Downs, *Manufacturing Inequality*, 31.

19. "An Employer's Protest," *County Advertiser*, 20 September 1913.

20. As discussed below, aside from the war years when the Kenricks and other employers in the hardware and hollow-ware trade were forced by the Ministry of Munitions to provide various workplace welfare benefits, these employers would continue to stall, resist, complain, and otherwise oppose virtually all efforts

by the Board of Trade to interfere with their right to manage their workplaces as they saw fit. And the evidence suggests the Kenricks led this resistance among the hollow-ware firms. See, for example, their role in the protracted negotiations with factory inspectors in the file, "Hollow-ware and Galvanized Welfare Order, 1921," PRO/LAB/14/203. An exasperated factory inspector, in reporting confidentially to his superior after the latest meeting, summarizes the attitudes of the Kenricks and the other employers by saying, "They had formed certain views of what the requirements were & nothing I said would change them. They took a stupid attitude," Letter from J. H. Walmsley to Chief Inspector, Office of H. M. Superintending Inspector of Factories, 18 January 1921. And finally, this contempt toward state intervention was expressed succinctly in a resolution passed by the Birmingham EEF on 13 February 1922 at a meeting of the Management Board: "Mr. Kenrick raised the question of the attitude of the Ministry of Labor in trying to define the classes of workpeople and work which should come under the various Trade Boards, particularly in conjunction with the stamped and pressed metal wares . . . It was resolved: That the Association oppose Trade Boards on principle, and suggest to members inquiring that they should inform the Ministry that Trade Boards were not necessary in their particular industry," "Minutes, February 13th, 1922," BDEA, MRC/MSS.265/B/1/5.

21. John Benson, *The Working Class in Britain*, 1850-1939 (London: Longman, 1989), 50-51.

22. "Developments in social legislation also strengthened the bargaining position of workers. A limited scheme of unemployment insurance, introduced in 1911, was slightly extended in 1916 and made general in 1920." See Norman Robertson and K. I. Sams, eds., *British Trade Unionism, Selected Documents* (Totowa, N. J.: Rowman & Littlefield, 1972), 1: 5.

23. DM, Thirtieth Annual Report, 1913.

24. Ministry of Labor and National Service, *Industrial Relations Handbook* (London: Her Majesty's Stationery Office, 1944), 17-19.

25. The agreement was signed by the National Union of Gas Workers, and the Amalgamated Workers, Brick Makers and General Laborers in addition to the Workers' Union, MRC/MSS.265/M/3/1; Askwith, *Industrial Problems and Disputes*, 252-258.

26. Askwith, *Industrial Problems and Disputes*, 256-257.

27. With respect to state intervention during this period Zeitlin writes, "From the 1890s onward, civil servants and politicians alike became committed to the extension of collective bargaining as a distinctly British method of reconciling economic efficiency and social peace. State intervention in industrial disputes reached its apogee under the Liberal administrations of 1906-1914, and the EEF became increasingly concerned about political constraints on employer unilateralism including the possible introduction of compulsory arbitration," Zeitlin, "The Triumph of Adversarial Bargaining," 414.

28. Church, *Kenricks in Hardware*, 146-147. According to the "Thirty-Second Annual Report": "Allowances are being made to many of those who have enlisted in the Army and Navy and their places will be kept open until the war has finished," DM, 14 August 1915. "Ministry of Munitions of War, Return From Controlled E al shments," 27 March 1916, Kenrick Collection.

29. An undated "press release" reads: "Some idea of the contribution of the Hardware Trade to the War Effort can be got from some figures recently released of Stores supplied by Archibald Kenrick & Sons Ltd. of West Bromwich. These figures also help to explain why some things have practically disappeared from the shops. The figures cover the period from the beginning of the War up to early this year: 117,000 pieces of cast iron hollow-ware; 44,000 large mincers; 20,000 small mincers; 3¼ Million pieces of enamelware; 3 Million Flag Clips in gun-metal or brass; 1¼ Million 2" Mortar Bombs; 2 Million Nose Containers for other Mortar Bombs; 7 Million Grenades." To this list, someone added in barely legible long-hand 3 more items in quantities of 400,000, 200,000, and 13 million. And in the DM for April 1919, it was reported that the Kenrick subsidiary, Anglo Enam-elware, had made for the War Office "3,133,156 Enameled Water Bottles charged at £344,167,178." Kenrick Collection.

30. DM, Thirty-second Annual Report, 1915.

31. DM, Thirty-third Annual Report, 1916.

32. DM, Thirty-fourth Annual Report, 1917.

33. DM, Thirty-fifth Annual Report, 1918.

34. Church, *Kenricks in Hardware*, 139, 147.

35. A "Mr. Kenrick," most likely either Clive Kenrick or Sir George Kenrick, was appointed to the committee established to press for changes in the procedures for computing and applying the excess profits tax. See "The Midland Employers' Federation Report of Special General Meeting, March 30th, 1916," MRC/MSS.265/M/3/2, and "The Midland Employers' Federation Report of the Annual General Meeting, January 4th, 1917," MRC/MSS.265/M/3/2.

36. "The Midland Employers' Federation Report of the Annual General Meeting, January 4th, 1917," MRC/MSS.265/M/3/2.

37. Church, *Kenricks in Hardware*, 296.

38. See G. E. Geobey, "Birmingham Area," Workers' Union *Record* no. 34 (August 1916): 5, MRC.

39. "List of Controlled Establishments," Vol IV, Section IV, item 54, Beveridge Collection on Munitions, British Library of Political and Economic Science.

40. Downs, *Manufacturing Inequality*, 33.

41. Ministry of Labor and National Service, *Industrial Relations Handbook*, 19-20.

42. "Minutes, October 23rd, 1916," BDEA, MRC/MSS.265/B/1/4.

43. "The Midland Employers' Federation Report of the Annual General Meeting, January 4th, 1917," BDEA, MRC/MSS.265/M/3/2.

44. "The Midland Employers' Federation Report of the Annual General Meeting, January 4th, 1917," BDEA, MRC/MSS.265/M/3/2.

45. "Presidential Address to First Triennial Conference by John Beard," Workers' Union *Record* no. 33 (July 1916): 5-6, MRC.

46. Similarly, the WU also appears to have been happy with the system. According to the notes of Richard Hyman, based on an interview conducted in the mid-1960s, S. Taylor, a WU organizer, remembered with some satisfaction that the: "Engineering employers already in Federation with full-time Secretary. No shop Stewards; workers with grievance would contact union secretary, who arranged meeting with Employers' Federation Secretary. Machinery worked well; avoided strikes." S. Taylor, interview by Richard Hyman, 20 September 1965, Richard Hyman Papers, MRC/MSS.51/3/1/48.

47. This seems to have been Askwith's view as well. See Askwith, *Industrial Problems and Disputes*, 257.

48. DM, Thirty-third Annual Report, 1916-1917; Church, *Kenricks in Hardware*, 297.

49. Downs, *Manufacturing Inequality*, 148-165, provides a useful discussion of the connection between welfare supervision and labor discipline which we draw on here.

50. Downs, *Manufacturing Inequality*, 158.

51. Downs, *Manufacturing Inequality*, 155.

52. Downs, *Manufacturing Inequality*, 158-159.

53. As quoted in Downs, *Manufacturing Inequality*, 155.

54. As quoted in Downs, *Manufacturing Inequality*, 151-152.

55. Downs, *Manufacturing Inequality*, 149-150.

56. "Executive Committee Meeting, April 22nd, 1918," BDEA, MRC/MSS.265/B/1/5.

57. "Return from Controlled Establishments, 7 November, 1918," Kenrick Collection.

58. "Controlled Establishments Processes in Which Women are Replacing Men," PRO/MUN/5/101/360/106.

59. "Dilution" did, apparently, create some problems for the WU, as we see in this letter from G. H. Oliver to Richard Hyman. Oliver was a member of the Amalgamated Society of Engineers (ASE) in 1910, eventually became Secretary of the Allied Engineering Trades, and was an organizer for the WU after 1920. He writes: "In reply to your letter concerning the relationship between the WU and other unions, particularly the ASE, there was very little if any friction during the years 1910-14. In these years the WU was concerned to organize the unskilled worker in all industries where the workpeople were unorganized. During the war years there was considerable dilution of labor in the Engineering Trade and many jobs hitherto performed by skilled men were 'broken down' into constituent parts and given to men and women with little or no knowledge of the trade. It was among this class of worker, the organizing of which produced friction between the

two unions. The re-graded men, many of whom were members of the WU, were eager to become members of a craft union, ASE, Coppersmiths, Sheet-Metal Members, Steam Engine Workers, brass and Metal Mechanics, and others . . . But in the main the ASE was the goal. It was a natural impulse for these men to get into a union consisting, in the main, of skilled workers. It was not so much a matter of poaching as a natural drift to improve their trade union status. The WU understandably resented the loss of members." Letter from G. H. Oliver to Richard Hyman, 8 December 1964, Richard Hyman Papers, MRC/MSS.51/3/1/39.

60. "Resolved, that employers will not hire workers from affiliated firms without the permission of that firm. It was proposed to cooperate with the Midland Employers' Federation in this regard." "Minutes of the Executive Committee Meeting, March 15th, 1915," BDEA, MRC/MSS.265/B/1/3.

61. DM, Thirty-third Annual Report, 1916.

62. In an interview with J. Townend, WU organizer, conducted in the mid-1960s, Hyman noted that in Townend's view, the WU supported the Munitions Act because it ". . . brought centralization of negotiation . . ." and in doing so, as we have seen, tended to strengthen the role of established unions and their leaders. Hyman also noted that, according to Townend, "To get good results out of the Act, Workers' Union pioneered trade Union education—most Employer's Representatives had no special training, easily bettered in negotiations." The latter might also explain why the WU supported the Act, and state intervention more generally. J. Townend, interview by Richard Hyman, 30 August 1965, Richard Hyman Papers MRC/MSS.51/3/1/51.

63. G. R. Askwith to the Ministry of Reconstruction, "Confidential Memorandum on Employers' Associations," 30 August 1917, PRO/RECO1/376,4.

64. Church, *Kenricks in Hardware*, 147.

65. "Draft Rules for an Association," 9 January 1918, Kenrick Collection.

66. In the papers of the Kenrick collection, undated, but obviously drawn up well after 1924, we have an attempt to analyze costs and productivity which suggests that between 1913 and 1920, the increases in wages the Kenricks had been paying were not accompanied by increased productivity. And while this might have not presented a problem to the firm during the war, when contracts with the MUN were generous and competition nonexistent, these costs were hurting the firm badly after the war. Calculations of wages cost per piece of "No. 500 latches" for example, show: in 1913 the firm got 41.5 pieces per pound of wages; in 1919 they got only 13.5 pieces per pound in wages; and in 1922 they got 15.5 pieces per pound in wages. On the bottom of the page someone had written the following comments: "Production per head was about 66 percent in 1919 of what it was in 1913. In 1920 production was about 90% of 1913 reduction due to waste and bad materials. In 1921 and on production seems to be at about pre-war rate. In 1924 55% of the number of workers working 12% shorter week could have turned out more than 55% of the 1913 output, if working full week. In 1924 (3 last months) we made 36 gross of 500 latches with 17 hands (3 males and 14 females) at wage

cost of £245, against an output of 46 gross for 6 months in 1914 with (6 males and 12 females) 18 hands at wage costs of £316, this is practically the same. In 1921 for 6 months we made 83 gross latches with 33 hands at cost of £1,084 or nearly double what we are now doing," "Report on Production of Cast Iron Hollow-ware and No. 500 Latches at Various Times," Kenrick Collection.

67. The "Allied Trades" employers were firms in the metal trades not considered to be, strictly speaking, engineering firms. They were in fact, with few exceptions, the 250 or so firms (of about 500 total firms in the Birmingham EEF) who constituted the MEF prior to amalgamation in 1918.

68. Clive provides his colleagues with progress reports at almost every bi-weekly meeting of the Birmingham EEF Management Board, beginning with the meeting on 30 November 1920, Minutes, BDEA, MRC/MSS.265/B/1/5. See also "Engineering Employers' West Midland Association. Base-Rate Negotiations with Workers' Union; Correspondence; Negotiating Conferences, 1898-1922," MRC/MSS.265/B/3/2.

69. For example, at the 8 February 1921 meeting of the Management Board we find in the minutes: "Mr. Clive Kenrick reported that a Local Conference had taken place February 4th, 1921 with regard to a reduction in wages, at which no agreement had been reached. It was therefore decided to refer the matter to York for Central Conference," MRC/MSS.265/B/1/6. This routine practice of referring disagreements to binding arbitration was one specific legacy of wartime state intervention, as indicated in a government publication in the 1940s: "This war-time national arbitration gave encouragement to the regulation of wages on a national basis, and after the war wage changes in many industries continued to be made on a national basis and through centralized organizations," Ministry of Labor and National Service, *Industrial Relations Handbook*, 19-20.

70. "Though it can never be satisfactory to accept a reduction of wages, there can be no doubt that the members of the Engineering Trade adopted the only sensible course left open to them when a ballot vote of all unions declared in favor of accepting a reduction rather than entering upon a strike. There are times when feelings must give way to reason, and such is the position today. None of us may be over impressed by the statement of the employers that they are unable to carry on and get trade at present contract prices, but we are bound to take notice of the other fact that directly concerns us, that large numbers of men have been unemployed for the past eight or nine months, and in the Engineering Trade it is quite safe to say that 50% of the work people have been more or less unemployed during that period," Workers' Union *Record* no. 93 (August 1921): 12, MRC.

71. With the war over, the Munitions Act was no longer in force. Hence, industrial relations reverted back to their pre-war, voluntary status.

72. "Mr. C. Kenrick reported that the following offer had been made to the representatives of the Societies concerned: The Employers are prepared to enter into an arrangement with you whereby the War advances to male workers in the Cast Iron Hollow-ware trade shall be regulated by reference to the index figure

with respect to the cost of living as set forth in the Board of Trade Labor Gazette, in the manner hereinafter provided," "Minutes, April 12th, 1921," BDEA, MRC/MSS.265/B/1/5.

73. See "Engineering Employers' West Midland Association. Base-rate Negotiations with Workers' Union; Correspondence; Negotiating Conferences, 1898-1922," MRC/MSS.265/B/3/2.

74. Zeitlin writes, "While a minority of firms manufacturing relatively standardized products pushed ahead with major changes in technology and management methods, the bulk of British engineering employers remained heavily dependent on skilled labor in the production process," "The Triumph of Adversarial Bargaining," 413-414.

75. Zeitlin, "The Triumph of Adversarial Bargaining," 412.

76. In the minutes of the 17 October 1921 meeting of the Management Board we find: "Mr. C. Kenrick, together with . . . reported that the Interview with Sir Allan Smith had taken place in London on Friday, Oct. 14th, 1921. At this interview the peculiar position of some of the Allied Trades Sections had been placed before Sir Allan Smith, who had expressed himself as appreciating the difficulty of the situation. The Chairman supplemented the remarks of the other members of the Board, but explained that Sir Allan Smith did not definitely state that he was agreed, in principle, to certain sections having complete freedom outside the authority of the Management Board. In view of the necessity of the matter being considered from all points of view, it was resolved: 'That a Sub-Committee should be elected, consisting of 4 members of the Allied Trade Section and four Members of the Engineering Trade to meet and discuss the matter.' Clive Kenrick was appointed to this committee," "Minutes, October 17th, 1921," BDEA, MRC/MSS.265/B/1/6.

77. "Minutes, December 12th, 1921," BDEA, MRC/MSS.265/B/1/6.

78. "Minutes, January 2nd, 1922," BDEA, MRC/MSS.265/B/1/6.

79. "Minutes, February 24th, 1922," BDEA, MRC/MSS.265/B/1/6.

80. Transcript of meeting between representatives of the Birmingham Allied Trades employers and Sir Allan Smith, 20 July 1922, "Allied Trades Employers' Association Reorganization, October, 1921-December, 1923," MRC/MSS.265/B/3/3.

81. "Minutes, May 1st, 1922," BDEA, MRC/MSS.265/B/1/6.

82. For example, at a Special Meeting of the Management Board on 1 May, we find the following entries: "Messrs. John Wright and Eagle Range Ltd. The Chairman reported that this Member was not posting the Notice in connection with the management of workpeople, and claimed that in accordance with an arrangement made with the late Midland Employer's Federation they were exempt from taking action of this kind. After discussion the matter was left in the hands of the Chairman. Messrs. F. H. Lloyd & Co. Ltd. The Chairman reported that this firm had refused to post the Notices, in the first instance, but had not signified their intention of following the Federations' instructions. Messrs. E. C. & J. Keay Ltd.

Difficulty was also being experienced in connection with this firm, and it was resolved: That Sir Harris Spencer should interview Mr. Keay," "Minutes, May 1st, 1922," BDEA, MRC/MSS.265/B/1/6.

83. "Minutes, May 8th, 1922," BDEA, MRC/MSS.265/B/1/6.

84. Directors of Archibald Kenrick and Sons Ltd. to Sir Allan Smith, Secretary of the Engineering Employer's Federation, London, 5 May 1922, Kenrick Collection.

85. The AEU was a "spent force" after this defeat, and would take many years to recover from it. See Zeitlin, "The Triumph of Adversarial Bargaining," 417. Before being allowed in the factories, workers were required to agree to the following:

1. The Employers have the right to manage their works and the Trade Unions the right to exercise the proper functions of trade Unions.
2. In the exercise of these rights
   (a) The parties shall have regard to provisions for Avoiding Disputes of 17th April, 1914, which are amplified by the Shop Stewards and Works Committee Agreement of 20th May, 1919, or such other procedure as may be agreed upon,
   (b) Notice should be given by the management to the workman concerned or their representatives of any material change in the recognized working conditions. The matter should thereupon be considered in accordance with the recognized procedure.

In the event of the workpeople or their representatives being unable to agree with the management on the proposal submitted, the management should be entitled to give a decision which should be observed pending the recognized procedure for discussion being gone through.

In all other questions the instructions of the management should be observed and discussion should follow the managerial act.

"Minutes, March 27th,1922," BDEA, MRC/MSS.265/B/1/6.

86. These meetings would continue well into 1923, but with inconclusive results. See the file entitled, "Allied Trades Employers' Association Reorganization, October, 1921-December, 1923," MRC/MSS.265/b/3/3.

87. Transcript of meeting between representatives of the Birmingham Allied Trades employers and Sir Allan Smith, 20 July 1922, "Allied Trades Employers' Association Reorganization, October, 1921-December, 1923," MRC/MSS.265/B/3/3.

88. "Minutes, June 8th, 1926," BDEA, MRC/MSS.265/B/1/6.

89. "Minutes, July 6th, 1926," BDEA, MRC/MSS.265/B/1/6.

90. In the British engineering trades, conflict may have been institutionalized but assumed an adversarial rather than cooperative character according to Jonathan Zeitlin in "The Triumph of Adversarial Bargaining."

91. Davies, "Inserting Gender into Burawoy's Theory of the Labour Process," 393-394.

92. Mary Lynn Stewart, *Women, Work, and the French State: Labour Protection and Social Patriarchy, 1879-1919* (Kingston, Ont.: McGill-Queen's University Press, 1989), 13. As Downs put it, "The 'welfare lady's' expert and intimate knowledge of the woman worker's capacities and character made her a valuable ally in management's wartime campaign to tighten discipline and raise output." Downs, *Manufacturing Inequality*, 147.

93. For a review see, Gillian Pascall, *Social Policy: A Feminist Analysis* (London: Tavistock Publications, 1986).

94. Dennis Smith, "Paternalism, Craft and Organizational Rationality," 214.

Chapter 6

1.   Burawoy, "Between the Labor Process and the State," 590.

2.   William Sites, "Primitive Globalization? State and Locale in Neoliberal Global Engagement," *Sociological Theory*, 18, no.1 (March 2000): 123. On globalization generally, see also David Held et al., *Global Transformations: Politics, Economics, and Culture* (Stanford, Calif.: Stanford University Press, 1999); Saskia Sassen, *Globalization and Its Discontents* (New York: The New Press, 1998); John Tomlinson, *Globalization and Culture* (Chicago: University of Chicago Press, 1999); Manuel Castells, *The Information Age: Economy, Society and Culture*, vol. 1, *The Rise of Network Society* (Oxford: Basil Blackwell, 1996); Peter Dicken, *Global Shift: Transforming the World Economy*, 3rd ed. (London: The Guilford Press, 1998); Anthony Giddens, *Runaway World: How Globalization Is Reshaping Our Lives* (New York: Routledge, 2000); John Gray, *False Dawn: The Delusions of Global Capitalism* (New York: The New Press, 1998); and Robert J. Antonio and Alessandro Bonanno, "A New Global Capitalism?: From 'Americanism and Fordism' to 'Americanization-Globalization,'" *American Studies Quarterly* (forthcoming).

3.   Sites, "Primitive Globalization?" 131.

4.   "The Betrayed; 80,000 in Protest March over Rover Axe," *The People,* <http://www.people.co.uk> (2 April 2000).

5.   Burawoy, "Between the Labor Process and the State," 603.

6.   "Blair rejects rescue plea from thousands of Longbridge workers," *AFX News Limited,* <http://www.afxnews.com> (3 April 2000).

7.   Mark Weisbrot, "Globalization for Whom?" *Cornell International Law Journal*, 31, no.3 (1998): 653.

8.   Sites, "Primitive Globalization?" 132.

9.   United Nations, *The World's Women 2000: Trends and Statistics* (New York: United Nations Publications, 2000). See also Saskia Sassen, *The Mobility of Labor and Capital: A Study in International Investment and Labor Flow* (Cambridge: Cambridge University Press, 1988); Kathryn B. Ward, *Women Workers and Global Restructuring* (Ithaca: Industrial and Labor Relations Press, 1990); Irene Tinker, *Persistent Inequalities: Women and World Development* (New York: Oxford University Press, 1990); Maria Mies, *Patriarchy and Accumulation on a World Scale: Women in the International Division of Labour* (London: Zed Books, 1986); Helen I. Safa, *The Myth of the Male Breadwinner: Women and Industrialization in the Caribbean* (Boulder: Westview Press, 1995); Edna Bonacich, *Global Production: The Apparel Industry in the Pacific Rim* (Philadelphia: Temple University Press, 1995); Wendy Chapkis and Cynthia Enloe, *Of Common Cloth: Women in the Global Textile Industry* (Amsterdam: Transnational Institute, 1983); and Aihwa Ong, *Spirits of Resistance and Capitalist Discipline: Factory Women in Malaysia* (Albany: State University of New York Press, 1987).

10. Susan H. Williams, "Globalization, Privatization, and a Feminist Public," *Indiana Journal of Global Studies* 4, no. 1 (1996): 99.

Appendix A

Children's Employment Commission (1862)
Third Report of the Commissioners
No. 3414-1 (1864)

Messrs. A. Kenrick and Sons, Ironfounders, West Bromwich.

716. Hollow ware, a name applied to saucepans and other kitchen articles, is made in these works, which are of high standing, as well as a variety of small articles in cast iron, *e.g.*, nails, hinges, &c. The greater part of the many boys are employed in the foundries, large but low buildings, the remainder in filing and turning. In a large shop full of steam-turned lathes, in which, however, there were no boys, I noticed one of the men with a respirator. I was told that these were supplied as a protection against the dust and fragments which fly off in considerable quantities, but that they were not much used.

717. When I entered the large foundry at 20 minutes to 2 it was fully at work, though only two-thirds of the dinner hour had passed. A young boy saying that a quarter of an hour was the time that he usually took for dinner, an elder boy, questioned by one of the employers if that was all, said that he thought that in the regular way they took about 20 or 25 minutes, not more.

718. *Mr. Kenrick.*—I do not think that there is anything in the employment of the young in the manufactures of this district which calls for legislative interference; and this is the opinion of those manufacturers in the district with whom I have spoken on the subject. I do not see however, that, so far as we ourselves are concerned, regulations of the factory kind would affect us injuriously. Our hours are from 6 to 6 with a half-day on Saturday, and for most of the hands on Monday also, and the regular hours of work are scarcely ever exceeded. Nearly the whole of the work depends upon the engine which blows the cupola, and this stops during meal-times. Some work, however can go on without it. If a man begins work as soon as he has finished eating, which perhaps he may do, as it is piece-work, the boys who help him must begin too. As many eat their meals in the works as away.

We do not wish for boys under 12 years old, but to allow only half time to all under 13 would throw a certain amount of loss on the families of those who are now employed. The practical effect of such a requirement in this district would be to shut out all under that age from work altogether, for they are generally employed by the workmen themselves, and these I am sure would never have to do with double sets. In our case all the boys, with one or two exceptions, are employed and paid by the workmen, who engage to complete so much work for a given price and find the labour.

If some employments were under regulation and others not, those which were not would have an advantage in the command of labour. Close to us there are very large glass works, as well as works of other kinds.

In one part of our work, which requires a furnace and goes at night, viz., the enamelling, three young boys from 11 or 12 upwards are employed for about nine or 10 hours in attending to the men. Their work is light, being merely to pull up balanced doors and to carry things to the furnace. This is work in which a man would not be of any more use than a boy.

It would be very difficult practically to carry out any regulations by inspection or otherwise in the smaller places, where there is more likelihood of their being needed, or indeed in the majority of the work-places, as so few work together. In many large places the boys are scattered about in ones and twos and threes in a shop under men, and there are a great number of small employers, "outworkers," who do not employ more than eight or 10 persons altogether, and work without machinery of any kind.

If any measure at all were thought necessary, the best would be to limit the age at which children could enter upon employments, say 12, but to leave them free after that to work the full time. Education previously to that age should not be enforced, though it is very desirable that it should be had. I consider that to enforce it, as *e.g.* in Prussia, would be objectionable in principle as an undue interference with private liberties. Even to require a test of some previous instruction on entry to employment is objectionable on the same ground. I was not aware of such a principle having been laid down by law in any case.

We established a school ourselves for the benefit of the children of our workpeople, at a weekly payment; but these did not fill it, so we allowed other persons to come. None who are in the works attend it. It is not open in the evening.

In this district there are but about half a dozen other hollow ware manufacturers, three in this parish, the others near Wolverhampton and Bilston. Three or four of the works may be about as large as our own. The system of work is probably the same as in ours.

719. *Charles Curley*, age 10—"Thread knuckles," *i.e.*, put parts of hinges together, "pun" dust, take up scrap (waste metal), take out sides (of moulds), riddle sand, skim metal, *i.e.*, take off the surface of the molten iron from the top of the pot just before it is poured, and clean the work. The other younger boys do much the same. Come at 6 a.m., or a little before, and leave at 6½ or 6¼ p.m. Meals in here. Breakfast at about 9, and begin work again as soon as it is done, usually about a quarter of an hour. About the same time for dinner. Get a wash every night at home. Another bigger boy works under the same man with me. Get 3*s*. 4*d*. a week.

Was at school six years till coming here 10 months ago, and paid 4*d*. a week, and go on Sunday, but never was at night school. Learned off the maps, reading, and sums. Can read. (Does fairly.) (Spells from sound "punished" and "America.") Can write. (Writes his name.) Four times 3 is 12.

720. *George Moore*, age 9—Am 10 next year, but do not know when. Work with father at the same work as the last boy.

Was never at school except on Sunday. Do not know B, O, or A; A is Y.

721. *Thomas Ferrars*, age 20—"Run sides," *i.e.*, put sand into moulds. Went to work at between 7 and 8 years old, blowing bellows and helping the man.

Go to school on Sunday, but never was at any other school. Cannot read, but know the letters (but calls "d" "b.") Father is not at work, having been asthmatical for eight years; was a collier. By what people say, London is a good place, and large. Have heard from a friend that Liverpool is a fine place. Have not heard of a mountain. Believe the Queen is a woman, but do not know what her name is, or if it is Victoria.

722. *James Ferrars*, age 17, brother of last witness—Went to school last Sunday. Did not before, because I had no clothes, and never was at any other school in my life, and have not been taught anything.

[These two brothers are said to have had great disadvantages at home, and to be irregular in their habits. Both are squalid and feeble looking; and the elder, though over 20, looks quite a boy, and so thin that every rib could be counted, the shirt being half gone. Indeed, the two shirts would scarcely make one between them. Some others were very ill clothed.]

723. *Joseph Baker*, age 16—Work at nail-casting. The apron is to keep the mould from hurting my legs and dress. Have been at work seven years.

Never at school except sometimes on Sunday. Cannot read. (Knows the letters.)

724. *George Jones*, age 13—File and turn at a lathe. Have been at these works since 7 years old. Get 7*s*. a week wages, and 3*d*. for sweeping for a woman.

Father cannot afford to pay for my going to night school. Can read a little (but imperfectly; *e.g.*, "was" is "as.")

725. *William Stringer*, age 12—Blow bellows. Pulling a rope with my foot and hands.

Cannot read. (Reads words of two or three letters, and writes from sound "boy" and "5.") Was at school till 10 years old.

726. *William Holden*, age 12—At school till 10 years old. Did multiplication. Five times 5 is 10, –is 20, –is 25. (Reads tolerably.) Am about the best scholar in this shop. There are nine boys in it. One can read as good as me.

Appendix B

Factory and Workshops Acts Commission
Report of the Commissioners Appointed To Inquire into the Working of the
Factory and Workshops Acts: With a View to Their Consolidation and
Amendment: Together with the Minutes of Evidence, Appendix, and Index.
Vol. 2. Minutes of Evidence
C 1443-I. 1876

John Arthur Kenrick, Esq., examined

6495. (*Chairman.*) You are an ironfounder and hollow ware manufacturer at West Bromwich, and also a manufacturer of some enamelled goods, into which the process of melting glass enters?—Yes.

6496. Does that fairly describe your branch of trade?—Strictly speaking we are ironfounders, and cast-iron hollow ware makers, and I represent not only my own firm, but the trade. I appear as the Chairman of the Hollow Ware Association. The enamel is the lining of culinary utensils with a sort of porcelain or glass instead of the ordinary tin.

6497. Incidental to your own business of ironfounding this trade enters?—Yes.

6498. Will you tell the Commission to what extent you employ women and children in all your works?—We employ about 87 women, and we have 35 half-timers, but the number of children between 13 and 18 I have not got, because I did not know that it would be interesting to the Commission, but I daresay there are from 140 to 150.

6499. As to those half-timers, how is their work arranged?—It is arranged in different ways. For some men they work mornings one week and afternoons the next week, and for others they work from 6 in the morning until dinner time, and then they send them to school in the afternoon.

6500. You mean that some of the children are at work every morning, and at school every afternoon?—Yes, and others work alternate weeks in the mornings and afternoons.

6501. Is that on account of the particular exigencies of the employment on which those children are engaged?—No, some of the men are rather wiser than others, and will allow the children to get fuller education by sending them alternate weeks than they would do if they worked in the mornings from 6 till 1.

6502. They think that it would be better for the children if they were obliged to be sent to school in the morning one week, and in the evening the next, or on alternate days?—The alternate days I should prefer very much indeed, that is to say, two days one week and three days the next week. They play on Saturday, and I do not know that there would be any objection to their working the half-day on Saturdays, because they only get into mischief if they are unemployed.

6503. Do your children generally work by shifts; would you have room for the whole of them on Saturday morning?—In some departments we might, but in most we could not.

6504. You would rather that it was left to your convenience whether the children worked on Saturday morning, or whether some of them were away for the whole day?—Yes, certainly.

6505. Will you tell the Commission in which of the processes of that work women are employed?—The use of women's labour has increased very much indeed since the beginning of the Factory Act; part of them are employed in tying up paper parcels of goods, small articles that are tied up by so many dozens; others are employed in japanning. They japan the outsides of saucepans and kettles, and in this work they have to follow the tinners. The tinners work so many hours, and the women follow them directly afterwards; and there I would say that the operation of the Factory Act has already deprived the men of something like two and a half hours' work, or a quarter day's work every week, because the men have to leave off work sooner to enable the women to leave off at the regulation time, that is at 6 o'clock. The women have to follow the men, and if they cannot finish in time, the men have to leave off sooner. In some of our other shops they are employed in doing light work with the hammer, it is not at all hard labour. Then again in another part they are employed in working light machines, cutting out covers from sheet iron; but there again the women either precede the operations of the men or follow on the work done by the men, and if their labour is curtailed practically it is a fine on the men, it shortens their hours. Nearly all our work is piece work, and the shortening of the men's hours is a very serious matter to them. None of our work is I should say hard work for the women.

6506. During what hours would the men work if they were not restrained by the women being taken away?—That would depend upon how many hours the women are allowed to work. They would work another two and a half hours to three hours a week but for the women leaving off.

6507. And they work up to what hour in the evening?—Up to 6. At present they have to work up to 5.30, except on one day, and then they have to leave off at 5.

6508. Why have they to stop work at 5 or 5.30?—Because the women japan their work after it is tinned, and we could not get it all japanned on the same day without the men leave off earlier.

6509. Would it meet your case if your work might be between 7 and 7?—No, that would not make the slightest difference, because we have the exact number of women that are required to do a certain number of men's work. They begin quite as soon or very nearly as soon as the men, but it is necessary that all this work should be japanned up the day on which it is done. It is a sort of continuous process.

6510. Do the men come in the morning earlier than the women?—Our hours are from 6 till 6, except on Mondays, when we begin at 7 and leave off at 5, and on Saturday we work from 6 till 1. We work 57½ hours a week instead of the ordinary number of 60.

6511. The men come at 6 in the morning, when do the women come?—They come at 6 too.

6512. I thought you said they came a little later?—Some of them do.

6513. Is it difficult with you to get women to come so early?—No, some years ago we used to work from 7 to 7, and it was for the convenience of the people that we altered it to 6 to 6. I think invariably that it is very much better for women to begin at 6 and leave off at 6 than it is for them to begin at 7 and work till 7.

6514. The Commission have been told that in this part of the country the men would not mind beginning earlier, but the women have a very great objection to it?—In Birmingham I believe it is so, but we employ very few married women, and it is better for unmarried women to go home as soon as possible in the daylight.

6515. What extension of time do you think is required not to interfere with your work?—My own feeling is, that take women at the same age as you do men, that is at 18, women should be allowed to work the same hours as men work and to have the privilege of working overtime if they like. I should put them on a perfect equality as regards making arrangements for the sale of their labour, whether married or single, and it appears to me that the more opportunities you give a woman of choosing her labour the better she is off.

6516. Is that true of her family too?—Yes, I think so.

6517. Then even if a woman has worked for 11 or 12 hours a day besides the intervals of labour, do you think that her family are likely to be as well attended to as if she was only working 9 or 10?—That depends upon circumstances. If there is a great demand for women's labour they will not work those number of hours unless they like. So far as my experience goes, if a woman does not like her occupation she gives notice and goes to somewhere else. This has been particularly the case lately. On account of the very great increase in the demand for their labour their wages have gone up from 20 to 40 per cent, within the last few years. And there is another circumstance which appears to me an advantage, that if a woman can get good wages for herself, in many cases she will remain single, she will not marry and be made to work for some of those characters whom we have heard described to-day. I know cases in which women have remained single, and are single at this present time, because they have been able to obtain sufficient wages to support themselves.

6518. I would ask you whether there is anything unhealthy in any of the processes of your manufacture?—I do not think there is.

6519. If it is not a secret of the trade, does arsenic enter into it?—No, there is no arsenic or lead, or any deleterious matter at all used in it. We are very carefull indeed to avoid even the semblance of anything of the sort.

6520. I suppose that although arsenic administered pure is injurious, it might in the process of manufacture do no harm to the vessel in which it is mixed?—If we were to put arsenic in, it might get amongst the food, and people would be taken ill, and if the Commission are going to look into that matter I would ask them to find out whether it is not used in the wrought-iron enamel. I am very certain it is, both arsenic and lead are very largely used in the wrought-iron enamel.

6521. In your enamel it is not?—No, not in ours; there is not the slightest trace of it; we have none in our works.

6522. I understand you desire to ask the Commission to recommend a certain privilege to you of working young persons at night?—Yes, in our enamelling process. I would say that before the Factory Act came into operation we used to work the boys about eight and a half hours a turn. It is a continuous process. The men work, say, from 10 in the morning till 4, and then leave off; and then from 4 to 10, and then from 10 to 4 in the morning, and again from 4 to 10.

6523. Then they have to do a little work besides?—They never used to work more than 50 hours a week. At the present time the boys, say, work from 6 in the morning till 6 at night; it is all day labour, but it is longer hours than they used to have to work before the Factory Act came into operation. We had at first a special exemption from the Secretary of State to allow us to go on working just as they do in Glass works and in Mills and Forges, where boys are employed to work at night, but for some reason or other that exemption was taken away.

6524. What are you obliged to do, as you cannot employ boys at night?—In the present case the men have been fortunate in finding a man who is partly an idiot, and he comes, and they have to give him very high wages to do this work at night. He can help them very little; he has not got the strength of a boy.

6525. What is the particular job that he has to do?—His general job is lifting up the oven door; we have a muffle which is heated red hot, and the work is put in until it gets red hot; then it is brought out, and we put some glass upon it, and before it gets cold it is put in again, and it is fused by the heat, and it has to go through this process three or four times. The only real work that he has to do is to keep pulling this door up and down, it is not hard work.

6526. Is that the only work for which you desire to keep boys at night?—That is the only work.

6527. In your business you work day and night, do you not?—We do in that branch only.

6528. (*O'Conor Don.*) Is this one of those special duties that you want boys to work at night for?—Yes.

6529. Then you only want one person for that?—You will understand that formerly the men used to work in sets, and every set had a boy, but now they have to combine together and pay this idiot between them. Before they used to work six hours at a time, and now this man works from 6 at night till 6 in the morning.

6530. He does the work for all the sets?—He does the work for the sets.

6531. (*Chairman.*) Have you heard any complaints of the work of the idiot?— The men grumble. Directly it was known that the Commission was coming down, the men came to me and said, we wish you could get this altered; we wish to go back to the old process. It is a very heavy fine upon the men.

6532. (*O'Conor Don.*) How many boys' places does he supply?—He would supply two, I suppose. It would be a great convenience to the men, and I do not think that it would hurt the boys in the slightest degree. They would always have some nights' rest, as we work them alternate weeks; one week they would always have good nights, and in the next week they would have to work part of the night.

6533. (*Mr. Knowles.*) That refers to your own works, where you employ men and boys?—Yes.

6534. Would the same apply to other works?—Yes, where they do the same business it would apply just the same.

6535. (*Chairman.*) But it would not pay to employ a man to do it?—No; that would be a shame. Men's labour is worth a great deal more than that ought to be.

6536. Is there any other point which you would like to mention to the Commission?—There are two points which I should like to mention. One is, that in regulating the hours, or making any recommendation as to the hours in which people should work, I should like the Commission to take into consideration the number of holidays which we are bound to give through the exigencies of our work. The holiday book shows from the 26th of last June up to this present time our workpeople played 18½ days. We take our stock about the 30th of June, and we have to play [sic] for repairs and whitewashing nearly 10 days. It took us 10 days last year.

6537. Do you whitewash, and so forth, with your own workmen?—We have done so; but our works have grown so big, that we asked a builder if he would do it for us this year, and he said he could not find the men; he could not get enough men to do it in time.

6538. Are your women and boys employed in whitewashing?—No, only the men.

6539. (*Mr. Knowles.*) Are you compelled by Act of Parliament to whitewash once a year?—Every 14 months.

6540. (*Chairman.*) I suppose your men would not like to do that work on the statutory holidays?—No. You see we have to stop all our machinery, and we could not arrange to do it in the statutory holidays at all. Our works have to stand whilst it is being done.

6541. The men would not do it in the Easter and Whitsun week, would they?— No, certainly not.

6542. Then you do not want to be obliged to keep the statutory holidays, seeing that you have to give so many involuntary holidays?—Yes, with the men all through South Staffordshire Easter has been as great a holiday as Whitsuntide is in Manchester and that district. Before the Act came in we always worked on Good Friday and Saturday. Now, we pay wages on the Thursday night, and the workpeople have to play Good Friday and Saturday and Easter Monday.

6543. The fact is that it is only the Good Friday that is the grievance?—Good Friday is the grievance, because it necessitates us to make them play on the Saturday too.

6544. That is the only point, and what you want is to be allowed to substitute Easter Monday for Good Friday?—Yes, we shall be quite contented to do that. There is one other point about the half-timers which I should like to mention to the Commission. The wages of half-timers have nearly doubled since the Factory Act came in, and of course all labour has gone up. We have had to increase our men's wages in some departments from 10 to 20 per cent in consequence of the great dearth of boy labour or the great prosperity of the country.

6545. Have you a complaint to make in consequence of that?—I do not wish to make any complaint if the hours are not shortened.

6546. Do you wish that the half-timers should have to work more than six and a half hours a day?—My wish is that the half-timers should be allowed to work the whole day on alternate days. I think that would be much the best for the school and much the best for the boys.

6547. Do you mean the whole day of 10½ hours?—Yes, 10½.

6548. You would object to its being made 10 hours?—Yes, decidedly; it would be a fine to the men because the men cannot work without the boys. There is also one other point which I should like to mention. I think that the hours of work ought to be allowed to be arranged, say, from 6 to 6 or 7 to 7 or 8 to 8, according to the requirements and pleasure of the contracting parties, that is to say, the workmen and the masters.

6549. It would be inconvenient, would it not, and lead to considerable difficulty of inspection if in one district there was a great difference of opinion and one place was working 6 to 6, and another 7 to 7, and another 8 to 8; seeing that many might be small works would not that lead to some difficulty?—It might lead to some difficulty, and it would throw upon the inspectors more trouble, but the inconvenience to the inspectors I think would be nothing in comparison to the convenience to the working people.

6550. You would like a choice to be allowed to the employer of working between 6 and 6, or 7 and 7, or 8 and 8?—Yes.

6551. (*Mr. Knowles.*) Would you compel each employer to give notice of his hours to the inspector?—Yes, certainly, there would not be the slightest objection to that.

6552. Whatever hours be adopted he should notify to the inspector?—Yes, we are obliged to put up in our works that our hours are from 6 to 6, and I think it is perfectly fair that the hours in which the works are going on should be stated.

6553. As regards the meal times, if you commence working at 6 what time do you allow for breakfast?—We allow half an hour for breakfast, from half-past 8 to 9, and then an hour for dinner, from 1 to 2.

6554. And then do you work them until 6 without a break?—Yes, but we do not exceed the four and a half hours at any time.

6555. But you only give them an hour and a half for meals?—Yes, I think that is quite enough; those are our extreme hours; sometimes they do not work so much as that.

6556. Can you get really 10½ hours' work when that is the case?—Yes, we do.

6557. In most trades where they commence at 6 they have half an hour for breakfast, and then one hour for dinner, and if they leave off at 6 they have half an hour for tea?—At one time when we worked from 7 to 7 the meal time was from 1 to 2, and the women used to have some time for tea, but that was abolished, and there is not the slightest practical difficulty in it. If they had not finished their dinner they would bring a bit of bread and cheese and eat it.

6558.   In the japanning process, has it to be continued until it leaves the hands of the workmen?—It is best; the only thing is this, that we have no room to store the work between the two processes.

6559.   But if you had room to take any accumulation that the workmen might have after the women ceased work could it be done the morning following?—Yes, but then the men begin to work directly in the morning.

6560.   If you employed more women you would do so?—We employ as many women as we possibly can; our shops are quite full now.

6561.   As regards the work that women have to do which precedes the men, would not the same apply if you had more women—could not they get up the morning's work for the men?—I was talking to one of our men about that very point, and he said that it was a very great inconvenience to him now when any of the women stopped away, it very seriously affected him.

6562.   Because he had nothing to go on with. But assuming that you had sufficient women to accumulate a little for him to go on with, so as to provide him with sufficient work to go on the whole time after the women ceased work, could you not manage it in that way?—Then two points are requisite; you would have to fine the manufacturer by making him provide more machines, and you would fine the workmen by making him employ more women.

6563.   Then he does not find sufficient women to keep him going?—Say six women can do the work, working from 6 to 6, if you take one hour off he may want seven women, and he would have to pay seven women at the same rate of wages that he paid the six. The manufacturer would have to pay so much more, and, of course, the public would have to pay so much more in the end.

6564.   (*O'Conor Don.*) Are the women paid by the piece or by the day?—Generally the women are paid by the day, but sometimes they are paid by the piece. May I be allowed to say with regard to half-timers that we find there is very great difficulty in getting certificates from the registrar. The charge now is 2*s.* 7*d.*, and for a boy who is getting 3*s.* a week that is a very heavy drawback. I think that as there really is no difficulty at all to the registrar to find out the boy's age the charge might be very materially reduced. The heavy charge was imposed upon the supposition that there was a very long search involved, but when a man is told the boy's age he can turn to the register and find out the place directly.

6565. What do you think would be a fair charge for a certificate?—I think that 6*d*. would be quite enough, and that would include the stamp; they have to put a penny stamp upon it. I certainly think that 3*d*. or 6*d*. or something of that sort would be quite enough.

6566. There is another point which I think is hardly necessary. For a boy to be working half-time the surgeon requires him to bring his certificate, whereas the surgeon is supposed to know by his teeth whether he is 13 or not, and the practical result of that is that you bring a half-timer up to your work, and you have taught him exactly what he ought to do, and when he becomes 13 the surgeon says, "Bring me your certificate," and the boy says, "No, I shall leave and go to another workshop and represent myself as 13." The surgeon looks at his mouth and says, "Oh yes, you are 13, you can work." If the expense of the certificate were reduced to some trifling sum, I do not think there would be any hardship at all.

6567. (*Chairman*.) Do you think that if the certificate was one lodged with the employer no further reference ought to be required?—I think that would be quite sufficient.

6568. (*Mr. Knowles*.) And if he changed his place of work it might be passed over to his next employer?—Yes, something of that sort would do very well indeed.

6569. What does the surgeon charge for the certificate?—The surgeon does not charge for the certificate, and I think he has no business to ask for the certificate; but he does.

6570. You employ no surgeon, as I understand?—We have to pay one under the Factory Act.

6571. When a boy obtained a certificate would not it be better for him to keep it as his own property as a passport to another place if he chose to leave?— I think it is very much better that the boy should keep it, but I saw it suggested by one of the inspectors that that might lead to fraud, and that he might hire it out or something of that sort, but I think that under all the circumstances I should make the owner, that is the boy who has paid for it, the keeper of it, and if he cheats he will be found out, and he ought to be punished.

The witness withdrew.

Select Bibliography

United Kingdom Archival Sources

Allied Trades Employers Association. Reports and Papers. At the Modern Records Centre [MRC], University of Warwick.

Amalgamated Engineering Union [AEU]. Reports and Papers. At the Modern Records Centre [MRC], University of Warwick.

Amalgamated Society of Engineers [ASE]. Reports and Papers. At the Modern Records Centre [MRC], University of Warwick.

Archibald Kenrick and Sons Ltd. Minutes of Meetings of Directors and Annual Reports [DM]. At Archibald Kenrick and Sons, Ltd., Union Street, West Bromwich.

Archibald Kenrick and Sons, Ltd. Papers [Kenrick Collection]. At the Black Country Museum, Dudley, West Midlands.

Beveridge Collection on Munitions. At the British Library of Political and Economic Science, London School of Economics and Political Science.

Birmingham District Engineering Trade Employers' Association [BDEA]. Minutes. At the Modern Records Centre [MRC], University of Warwick.

Birmingham Hospital Saturday Fund. Annual Report of the Birmingham Hospital Saturday Fund, 1883 and after. At the Birmingham Public Library.

Birmingham Trades Council. Executive Committee. Minutes. At the Birmingham Central Library, Chamberlain Square, Birmingham.

Cast Iron Butt Hinge Makers Association. Minutes. At the Black Country Museum, Dudley, West Midlands.

Cast Iron Hollow-Ware Makers Association [CIHMA]. Minutes. At the Black Country Museum, Dudley, West Midlands.

Engineering and National Employers' Federation, Birmingham and Wolverhampton District Association. Minutes. At the Modern Records Centre [MRC], University of Warwick.

Gertrude Tuckwell Collection (Microfilm). At the Center for British Studies, University of Colorado.

Julia Varley Papers. At the University of Hull.

Midland Employers' Federation [MEF]. Reports and Papers. At the Modern Records Centre [MRC], University of Warwick.

Piercy Papers. At the British Library of Political and Economic Science, London School of Economics and Political Science.

National Employers' Federation [EEF]. Reports and Papers. At the Modern Records Centre [MRC], University of Warwick.

Richard Hyman Papers. At the Modern Records Centre [MRC], University of Warwick.

T&C Clark Collection. At the Wolverhampton Borough Archives and Local Studies, Wolverhampton.

Workers' Union [WU], *Record*. At the Modern Records Centre [MRC], University of Warwick.

United Kingdom Government Documents

Census

1841-1891, Enumerators' Schedule of Households in West Bromwich. At the Local Studies Centre, Smethwick Library, High Street, Smethwick.

Parliamentary Papers

United Kingdom. Parliament. Children's Employment Commission (1862) [CEC]. *Third Report of the Commissioners*. No. 3414-1. 1864.

United Kingdom. Parliament. Factory and Workshops Acts Commission. *Report of the Commissioners Appointed To Inquire into the Working of the Factory and Workshops Acts: With a View to Their Consolidation and Amendment: Together with the Minutes of Evidence, Appendix, and Index*. Vol. 2. *Minutes of Evidence*. C 1443-I. 1876.

United Kingdom. Parliament. *Reports of the Inspectors of Factories to Her Majesty's Principal Secretary of State for the Home Department for the Half Year Ending 31st October 1868*. No. 4093-I. 1869.

Other

Ministry of Labor [LAB]. Records. At the Public Record Office [PRO], Kew.
Ministry of Labor [LAB]. Trade Boards Acts, 1909 and 1918, Hollow-Ware Trade. At the Public Record Office [PRO], Kew.
Ministry of Labor and National Service. *Industrial Relations Handbook*. London: Her Majesty's Stationery Office, 1944.
Ministry of Munitions [MUN]. Records. At the Public Record Office [PRO], Kew.
Ministry of Reconstruction [RECO]. Records. At the Public Record Office [PRO], Kew.
Office of the Commissioners of Patents (Frederick Ryland, various).
Report to the Local Government Board on the Sanitary Condition of the Urban Sanitary District of West Bromwich. London: Her Majesty's Stationery Office, 1875.

United Kingdom Newspapers and Periodicals

Aris Gazette; County Advertiser; West Bromwich Weekly News; The Birmingham Gazette; The Birmingham Daily Gazette; Midland Chronicle and Free Press; The Midland Chronicle for West Bromwich and Oldbury; The Free Press; The Wednsbury Boroughs News and Darlaston Chronicle; The Journal; The Sunday Chronicle; The Times; Manchester Guardian; Birmingham Mail; Woverhampton Chronicle; Old West Bromwich; The Ironmonger; The Metal Worker; The Hardware Trades Journal; The Foundry Trades Journal; The Iron Age.

Books and Articles

Aitken, W. C. "Brass and Brass Manufacturers." In *Birmingham and the Midland Hardware District*, edited by British Association for the Advancement of Science and Samuel Timmins, 225-380. London: Cass, 1967 [1866].
Alexander, Sally. *Women's Work in Nineteenth-Century London: A Study of the Years 1820-50*. London: Journeyman Press and London History Workshop Centre, 1983.
Allen, G. C. *The Industrial Development of Birmingham and the Black Country, 1860-1927*. London: Allen and Unwin, 1929.
Antonio, Robert J., and Alessandro Bonanno. "A New Global Capitalism?: From 'Americanism and Fordism' to 'Americanization-Globalization.'" *American Studies Quarterly*. Forthcoming.
Askwith, George. *Industrial Problems and Disputes*. London: John Murray, 1920.

Barnsby, George J. *Social Conditions in the Black Country, 1800-1900.* Wolverhampton, England: Integrated Publishing Services, 1980.

Baron, Ava, ed. *Work Engendered: Toward a New History of American Labor.* Ithaca: Cornell University Press, 1991.

Barrett, Michelle, and Mary McIntosh. "The 'Family Wage': Some Problems for Socialists and Feminists." *Capital and Class* 11 (summer 1980): 51-72.

Beechey, Veronica. "On Patriarchy." *Feminist Review* 3 (1979): 66-83.

Behagg, Clive. *Politics and Production in the Early Nineteenth Century.* London: Routledge, 1990.

Benenson, Harold. "The Family Wage and Working Women's Consciousness in Britain, 1880-1914." *Politics and Society* 19, no. 1 (1991): 71-108.

Benson, John. *The Working Class in Britain, 1850-1939.* London: Longman, 1989.

Berg, Maxine. *The Age of Manufactures: Industry, Innovation and Work in Britain, 1700-1820.* 2nd ed. New York: Routledge, 1994.

———. "Women's Work, Mechanization, and the Early Phases of Industrialization in England." In *The Historical Meanings of Work*, edited by Patrick Joyce, 64-98. Cambridge: Cambridge University Press, 1987.

Berlanstein, Lenard R., ed. *Rethinking Labor History: Essays on Discourse and Class Analysis.* Urbana: University of Illinois Press, 1993.

"Bibliography: Women and Work." *Journal of Women's History* 1, no. 1 (spring 1989): 138-169.

Bonacich, Edna. *Global Production: The Apparel Industry in the Pacific Rim.* Philadelphia: Temple University Press, 1995.

Bonnell, Victoria E. *Roots of Rebellion: Workers' Politics and Organizations in St. Petersburg and Moscow, 1900-1914.* Berkeley: University of California Press, 1983.

Boris, Eileen. "Beyond Dichotomy: Recent Books in North American Women's Labour History." *Journal of Women's History* 4, no. 3 (winter 1993): 162-179.

Boston, Sarah. *Women Workers and the Trade Union Movement.* London: Davis-Poynter, 1980.

Braverman, Harry. *Labor and Monopoly Capital: The Degradation of Work in the Twentieth Century.* New York: Monthly Review Press, 1975.

Braybon, Gail. *Women Workers in the First World War: The British Experience.* London: Croom Helm, 1981.

British Association for the Advancement of Science, and Samuel Timmins. *Birmingham and the Midland Hardware District.* London: Cass, 1967 [1866].

Brown, E. H. Phelps. *The Growth of British Industrial Relations: A Study from the Standpoint of 1906-1914.* London: Macmillan, 1965.

Burawoy, Michael. "Between the Labor Process and the State: The Changing Face of Factory Regimes under Advanced Capitalism." *American Sociological Review* 48, no. 5 (October 1983): 587-605.

———. "Karl Marx and the Satanic Mills: Factory Politics under Early Capitalism in England, the United States, and Russia." *American Journal of Sociology* 90, no. 2 (September 1984): 247-282.

———. *Manufacturing Consent: Changes in the Labor Process under Monopoly Capitalism*. Chicago: University of Chicago Press, 1979.

———. *The Politics of Production: Factory Regimes under Capitalism and Socialism*. London: Verso, 1985.

Cadbury, Edward, M. Cecile Matheson, and George Shann. *Women's Work and Wages: A Phase of Life in an Industrial City*. Chicago: University of Chicago Press, 1907.

Castells, Manuel. *The Information Age: Economy, Society and Culture:* Volume 1. *The Rise of Network Society*. Oxford: Basil Blackwell, 1996.

Chapkis, Wendy, and Cynthia Enloe. *Of Common Cloth: Women in the Global Textile Industry*. Amsterdam: Transnational Institute, 1983.

Church, R. A. "Family and Failure: Archibald Kenrick and Sons, Ltd., 1900-1950." In *Essays in British Business History*, edited by Barry Supple, 103-123. Oxford: Clarendon Press, 1977.

———. *Kenricks in Hardware: A Family Business, 1791-1966*. Newton Abbot, England: David and Charles, 1969.

Clegg, Hugh Armstrong. *A History of British Trade Unions since 1889*. Vol. 2: *1911-1933*. Oxford: Clarendon Press, 1985.

Corbett, John. *The Birmingham Trades Council, 1866-1966*. London: Lawrence and Wishart, 1966.

Court, W. H. B. *The Rise of the Midland Industries, 1600-1838*. 2nd ed. London: Oxford University Press, 1953 [1938].

Cressey, P., and J. MacInnes. "Voting for Ford: Industrial Democracy and the Control of Labour." *Capital and Class* 11 (summer 1980): 5-33.

Dangerfield, George. *The Strange Death of Liberal England*. New York: Capricorn Books, 1961 [1935].

Davies, Scott. "Inserting Gender into Burawoy's Theory of the Labour Process." *Work, Employment & Society* 4, no. 3 (September 1990): 391-406.

Dicken, Peter. *Global Shift: Transforming the World Economy*. 3rd ed. London: The Guilford Press, 1998.

Dilworth, D. *West Bromwich before the Industrial Revolution*. Tipton, England: Black Country Society, 1973.

Downs, Laura Lee. *Manufacturing Inequality: Gender Division in the French and British Metalworking Industries, 1914-1939*. Ithaca: Cornell University Press, 1995.

Drake, Barbara. *Women in the Engineering Trades*. London: Fabian Research Department, 1917.

———. *Women in Trade Unions*. London: Virago Press, 1984 [1920].

Edwards, Richard. *Contested Terrain: The Transformation of the Workplace in the Twentieth Century*. New York: Basic Books, 1979.

Eley, Geoff. "Edward Thompson, Social History and Political Culture: The Making of a Working-Class Public, 1780-1850." In *E. P. Thompson: Critical Perspectives*, eds. Harvey J. Kaye and Keith McClelland (Philadelphia: Temple University Press, 1990).

Engels, Friedrich. *The Condition of the Working Class in England*. Stanford, Calif.: Stanford University Press, 1968 [1844].

Frader, Laura L., and Sonya O. Rose, eds. *Gender and Class in Modern Europe*. Ithaca, N.Y.: Cornell University Press, 1996.

Friedman, Andrew L. *Industry and Labour: Class Struggle at Work and Monopoly Capitalism*. London: Macmillan, 1977.

Frost, P. M. "The Growth and Localization of Rural Industry in South Staffordshire, 1560-1720." Ph.D. diss., University of Birmingham, 1973.

Gannage, C. "A World of Difference: The Case of Women Workers in a Canadien Garment Factory." In *Feminism and Political Economy: Women's Work, Women's Struggles*, edited by H. Maroney and M. Luxton, 139-165. Toronto: Meuthuen, 1987.

Giddens, Anthony. *Central Problems in Social Theory: Action, Structure, and Contradiction in Social Analysis*. Berkeley: University of California Press, 1979.

———. *Runaway World: How Globalization Is Reshaping Our Lives*. New York: Routledge, 2000.

Gill, Conrad, and Asa Briggs. *History of Birmingham*. London: Oxford University Press, 1952.

Gordon, David M., Richard Edwards, and Michael Reich. *Segmented Work, Divided Workers: The Historical Transformation of Labor in the United States*. Cambridge: Cambridge University Press, 1982.

Gray, Jane. "Gender and Uneven Working-Class Formation in the Irish Linen Industry." In *Gender and Class in Modern Europe*, edited by Laura L. Frader and Sonya O. Rose, 37-56. Ithaca: Cornell University Press, 1996.

Gray, John. *False Dawn: The Delusions of Global Capitalism*. New York: The New Press, 1998.

Hanagan, Michael P. *The Logic of Solidarity: Artisans and Industrial Workers in Three French Towns, 1871-1914*. Urbana: University of Illinois Press, 1980.

Hannah, Leslie. *The Rise of the Corporate Economy: The British Experience*. Baltimore: Johns Hopkins University Press, 1976.

Harrison, Brian. *Separate Spheres: The Opposition to Women's Suffrage in Britain*. New York: Holmes and Meier, 1978.

Hartmann, Heidi. "Capitalism, Patriarchy, and Job Segregation by Sex." In *Women and the Workplace: The Implications of Occupational Segregation*, edited by Martha Blaxall and Barbara Reagan, 137-169. Chicago: University of Chicago Press, 1976.

———. "The Unhappy Marriage of Marxism and Feminism: Towards a More Progressive Union." *Capital and Class* 8 (summer 1979): 1-33.

Haydu, Jeffrey. *Between Craft and Class: Skilled Workers and Factory Politics in the United States and Britain, 1890-1922*. Berkeley: University of California Press, 1988.

Held, David, Anthony McGrew, David Goldblatt, and Jonathan Perraton. *Global Transformations: Politics, Economics and Culture*. Stanford, Calif.: Stanford University Press, 1999.

Heron, Craig, and Robert H. Storey. *On the Job: Confronting the Labour Process in Canada*. Kingston, Ont.: McGill-Queen's University Press, 1986.

Hobsbawm, Eric J. *Labouring Men: Studies in the History of Labour*. New York: Basic Books, 1964.

———. *Workers: Worlds of Labor*. New York: Pantheon Books, 1984.

Hunt, E. H. *British Labour History, 1815-1914*. Atlantic Highlands, N.J.: Humanities Press, 1981.

Hyman, Richard. *The Workers' Union*. Oxford: Clarendon Press, 1971.

Jefferys, James B. *The Story of the Engineers, 1800-1945*. London: Lawrence and Wishart, 1946.

Jephcott W. E. *The House of Izon: The History of a Pioneer Firm of Ironfounders*. London: Murray-Watson, 1948.

John, Angela V., ed. *Unequal Opportunities: Women's Employment in England 1800-1918*. Oxford: Basil Blackwell, 1985.

Jones, Edgar. *A History of GKN*. Vol. 1: *Innovation and Enterprise, 1759-1918*. London: Macmillan, 1987.

Jones, Gareth Stedman. *Languages of Class: Studies in English Working Class History, 1832-1982*. Cambridge: Cambridge University Press, 1983.

Joyce, Patrick. "The Historical Meanings of Work: An Introduction." In *The Historical Meanings of Work*, edited by Patrick Joyce, 1-30. Cambridge: Cambridge University Press, 1987.

———. *Work, Society and Politics: The Culture of the Factory in Later Victorian England*. New Brunswick, N.J.: Rutgers University Press, 1980.

Katznelson, Ira, and Aristide R. Zolberg. *Working-Class Formation: Nineteenth-Century Patterns in Western Europe and the United States*. Princeton, N.J.: Princeton University Press, 1986.

Kenrick, William. "The Hollow-Ware Trade." In *Birmingham and the Midland Hardware District*, edited by British Association for the Advancement of Science and Samuel Timmins, 103-109. London: Cass, 1967 [1866].

Knights, David, and Hugh Willmott. "Power and Subjectivity at Work: From Degradation to Subjugation in Social Relations." *Sociology* 23, no. 4 (1989): 535-558.

Kriedte, Peter, Hans Medick, and Jürgen Schlumbohm, eds. *Industrialization before Industrialization: Rural Industry in the Genesis of Capitalism*. Cambridge: Cambridge University Press, 1981.

Landes, David S. *The Unbound Prometheus: Technological Change and Industrial Development in Western Europe from 1750 to the Present*. Cambridge: Cambridge University Press, 1969.

Leask, James, and Philomena Bellars. *"Nor Shall the Sword Sleep . . .": An Account of Industrial Struggle*. Copy in the Birmingham Public Library: N.p., 1954.

Lee, Ching Kwan. "Engendering the Worlds of Labor: Women Workers, Labor Markets, and Production Politics in the South China Economic Miracle." *American Sociological Review* 60, no. 3 (June 1995): 378-397.

———. "Familial Hegemony: Gender and Production Politics on Hong Kong's Electronic Shopfloor." *Gender and Society* 7, no. 4 (1993): 529-547.

Lewenhak, Sheila. *Women and Trade Unions: An Outline History of Women in the British Trade Union Movement*. New York: St. Martin's Press, 1977.

Lewis, Jane. *Labour and Love: Women's Experience of Home and Family, 1850-1940*. Oxford: Basil Blackwell, 1986.

Littler, Craig R. *The Development of the Labour Process in Capitalist Societies: A Comparative Study of the Transformation of Work Organization in Britain, Japan, and the USA*. London: Heinemann, 1982.

Littler, Craig R., and G. Salaman. "Bravermania and Beyond: Recent Theories of the Labour Process." *Sociology* 16, no. 2 (May 1982): 251-269.

Lown, Judy. "Not So Much a Factory, More a Form of Patriarchy: Gender and Class During Industrialization." In *Gender, Class and Work*, edited by Eva Gamarnikow, David H. J. Morgan, June Purvis, and Daphne E. Taylorson et al., 28-45. London: Heinemann, 1983.

———. *Women and Industrialization: Gender at Work in Nineteenth-Century England*. Minneapolis: University of Minnesota Press, 1990.

Mappen, Ellen. "Strategists for Change: Social Feminist Approaches to the Problems of Women's Work." In *Unequal Opportunities: Women's Employment in England 1800-1918*, edited by Angela V. John, 236-259. Oxford: Basil Blackwell, 1985.

Marx, Karl. *Capital*. Vol. 1. New York: Vintage Books, 1977 [1867].

Marx, Karl, and Frederich Engels. *The Communist Manifesto: A Modern Edition*. London: Verso, 1998 [1848].

Meacham, Standish. *A Life Apart: The English Working Class, 1890-1914*. Cambridge, Mass.: Harvard University Press, 1977.

———. "The Sense of an Impending Clash: English Working-Class Unrest before the First World War." *American Historical Review* 77, no. 5 (December 1972): 1343-1364.

Mendels, Franklin. "Proto-Industrialization: The First Phase of the Industrialization Process." *Journal of Economic History* 32, no. 1 (1972): 241-261.

Mies, Maria. *Patriarchy and Accumulation on a World Scale: Women in the International Division of Labour*. London: Zed Books, 1986.

Milkman, Ruth. "Organizing the Sexual Division of Labor: Historical Perspectives on 'Women's Work' and the American Labor Movement." *Socialist Review* (January-February 1980): 105-108.

Noble, David F. *America by Design: Science, Technology, and the Rise of Corporate Capitalism*. Oxford: Oxford University Press, 1977.

———. "Social Choice in Machine Design: The Case of Automatically Controlled Machine Tools." In *Case Studies in the Labor Process*, edited by Andrew S. Zimbalist, 18-50. New York: Monthly Review Press, 1979.

Ong, Aihwa. *Spirits of Resistance and Capitalist Discipline: Factory Women in Malaysia*. Albany: State University of New York Press, 1987.

Parr, Joy. *The Gender of Breadwinners: Women, Men, and Change in Two Industrial Towns, 1880-1950*. Toronto: University of Toronto Press, 1990.

Pascall, Gillian. *Social Policy: A Feminist Analysis*. London: Tavistock Publications, 1986.

Pelling, Henry M. "The Knights of Labour in Britain." *Economic History Review* 9, no. 2 (December 1956): 313-331.

Pinchbeck, Ivy. *Women Workers and the Industrial Revolution, 1750-1850*. London: Cass, 1969 [1930].

Pollard, Sidney, and Paul Robertson. *The British Shipbuilding Industry, 1870-1914*. Cambridge, Mass.: Harvard University Press, 1979.

Reid, Douglas. "The Decline of Saint Monday, 1766-1876." *Past and Present* 71 (May 1976): 76-101.

Roberts, Elizabeth, and Economic History Society. *Women's Work 1840-1940*. Basingstoke: Macmillan, 1988.

Robertson, Norman, and K. I. Sams, eds. *British Trade Unionism, Selected Documents*. Vol. 1. Totowa, N.J.: Rowman and Littlefield, 1972.

Rose, Sonya O. "Gender Antagonism and Class Conflict: Exclusionary Strategies of Male Trade Unionists in Nineteenth-Century Britain." *Social History* 13, no. 2 (1988): 191-208.

———. "Gender Segregation in the Transition to the Factory: The English Hosiery Industry, 1850-1910." *Feminist Studies* 13, no. 1 (spring 1987): 163-184.

———. *Limited Livelihoods: Gender and Class in Nineteenth-Century England*. Berkeley: University of California Press, 1992.

———. "'Gender at Work': Sex, Class and Industrial Capitalism." *History Workshop Journal: A Journal of Socialist Historians* 21 (1986): 113-131.

Rowlands, Marie B. *Masters and Men in the West Midland Metalware Trades before the Industrial Revolution*. Manchester: Manchester University Press, 1975.

Safa, Helen I. *The Myth of the Male Breadwinner: Women and Industrialization in the Caribbean*. Boulder: Westview Press, 1995.

Sassen, Saskia. *Globalization and Its Discontents*. New York: The New Press, 1998.

————. *The Mobility of Labor and Capital: A Study in International Investment and Labor Flow*. Cambridge: Cambridge University Press, 1988.

Scott, Joan Wallach. *The Glassworkers of Carmaux: French Craftsmen and Political Action in a Nineteenth-Century City*. Cambridge, Mass.: Harvard University Press, 1974.

Seccombe, Wally. "Patriarchy Stabilized: The Construction of the Male Breadwinner Wage Norm in Nineteenth-Century Britain." *Social History* 2, no.1 (January 1986): 53-76.

Sewell, William H. Jr. "Toward a Post-Materialist Rhetoric for Labor History." In *Rethinking Labor History: Essays on Discourse and Class Analysis*, edited by Lenard R. Berlanstein, 15-38. Urbana: University of Illinois Press, 1993.

Sites, William. "Primitive Globalization? State and Locale in Neoliberal Global Engagement." *Sociological Theory* 18, no. 1 (March 2000): 121-144.

Smith, Dennis. *Conflict and Compromise: Class Formation in English Society, 1830-1914: A Comparative Study of Birmingham and Sheffield*. London: Routledge and Kegan Paul, 1982.

————. "Paternalism, Craft and Organizational Rationality 1830-1930: An Exploratory Model." *Urban History* 19, pt. 2 (October 1992): 211-228.

Smith, William Hawkes. *Birmingham and Its Vicinity as a Manufacturing District*. London: Charles Tilt, 1836.

Soldon, Norbert C. *Women in British Trade Unions, 1874-1976*. Totowa, N.J.: Rowman and Littlefield, 1978.

Staples, Clifford L., and William G. Staples. "'A Strike of Girls': Gender and Production Politics in the British Metal Trades, 1913." *Journal of Historical Sociology* 12, no. 2 (June 1999): 158-180.

Staples, William G. "Technology, Control, and the Social Organization of Work at a British Hardware Firm, 1791-1891." *American Journal of Sociology* 93, no. 1 (July 1987): 62-88.

Stewart, Mary Lynn. *Women, Work, and the French State: Labour Protection and Social Patriarchy, 1879-1919*. Kingston, Ont.: McGill-Queen's University Press, 1989.

Storper, M., and R. Walker. "Theory of Labour and the Theory of Location." *International Journal of Urban and Regional Research* 7, no. 1 (1983): 1-44.

Taylor, Arthur J. "The Economy." In *Edwardian England 1901-1914*, edited by Simon Nowell-Smith, 103-138. London: Oxford University Press, 1964.

Thom, Deborah. "The Bundle of Sticks: Women, Trade Unionists and Collective Organization before 1918." In *Unequal Opportunities: Women's Employment in England 1800-1918*, edited by Angela V. John, 261-289. Oxford: Basil Blackwell, 1985.

————. "Women at the Woolwich Arsenal 1915-1919." *Oral History* 6, no. 2 (1978): 58-73.

Thompson, E. P. *The Making of the English Working Class*. New York: Vintage Books, 1963.

Tickner, Lisa. *The Spectacle of Women: Imagery of the Suffrage Campaign, 1907-14.* Chicago: University of Chicago Press, 1988.

Tilly, Charles. *Durable Inequality.* Berkeley: University of California Press, 1998.

Tilly, Louise, and Joan Wallach Scott. *Women, Work, and Family.* New York: Holt, Rinehart and Winston, 1978.

Tinker, Irene. *Persistent Inequalities: Women and World Development.* New York: Oxford University Press, 1990.

Tomlinson, John. *Globalization and Culture.* Chicago: University of Chicago Press, 1999.

Trainor, Richard H. *Black Country Elites: The Exercise of Authority in an Industrialized Area, 1830-1900.* Oxford: Clarendon Press, 1993.

Turbin, Carole. *Working Women of Collar City: Gender, Class, and Community in Troy, New York, 1864-86.* Urbana: University of Illinois Press, 1992.

United Nations. *The World's Women, 2000: Trends and Statistics.* 3rd ed. Social Statistics and Indicators. New York: United Nations, 2000.

Walby, Sylvia. *Patriarchy at Work: Patriarchal and Capitalist Relations in Employment.* Cambridge: Polity Press, 1986.

———. *Theorizing Patriarchy.* Oxford: Basil Blackwell, 1990.

Wallace, Michael, and Arnie Kallenberg. "Industrial Transformation and the Decline of Craft: The Decomposition of Skill in the Printing Industry, 1931-1978." *American Sociological Review* 47, no. 3 (June 1982): 307-324.

Ward, Kathryn B. *Women Workers and Global Restructuring.* Ithaca, N.Y.: Industrial and Labor Relations Press, 1990.

Warde, Alan. "Industrial Discipline: Factory Regime and Politics in Lancaster." *Work, Employment & Society* 3, no. 1 (1989): 49-63.

Weber, Max. *The Protestant Ethic and the Spirit of Capitalism.* New York: Scribner and Sons, 1958 [1920].

Weisbrot, Mark. "Globalization for Whom?" *Cornell International Law Journal* 31, no. 3 (1998): 631-655.

Williams, Susan H. "Globalization, Privatization, and a Feminist Public." *Indiana Journal of Global Studies* 4, no. 1 (1996): 97-107.

Woodall, Richard D. *West Bromwich Yesterdays: A Short Historical Study of "The City of a Hundred Trades."* Sutton, England: Norman A.Tector, 1958.

Woods, D. C. "The Operation of the Master and Servants Act in the Black Country, 1858-1875." *Midland History* 7 (1982): 93-115.

Zeitlin, Jonathan. "The Internal Politics of Employer Organization: The Engineering Employers' Federation 1896-1939." In *The Power To Manage? Employers and Industrial Relations in Comparative Perspective*, edited by Steven Tolliday and Jonathan Zeitlin, 53-79. London: Routledge, 1991.

———. "The Labour Strategies of British Engineering Employers, 1890-1922." In *Managerial Strategies and Industrial Relations: An Historical Comparative Study*, edited by Howard F. Gospel and Craig R. Littler, 25-54. London: Heinemann, 1983.

————. "The Triumph of Adversarial Bargaining: Industrial Relations in British Engineering, 1880-1939." *Politics and Society* 18, no. 3 (1990): 405-426.

Zimbalist, Andrew. "Technology and the Labor Process in the Printing Industry." In *Case Studies in the Labor Process*, edited by Andrew Zimbalist, 103-126. New York: Monthly Review Press, 1979.

Index

AEU. *See* Amalgamated Engineering Union (AEU)

Aitken, W. C., 31, 139n2

Alexander, Sally, 145n14

Allen, G. C., 134n4, 134n7, 137n33, 137n37, 138n48, 139n19, 142n70

Allied Trades, 111, 113–18, 156n6, 162n67, 163n76, 163n80, 164nn86–87. *See also* Engineering Employers' Federation (EEF)

Amalgamated Engineering Union (AEU), 114–15, 117, 118, 154n118, 156n10, 164n85. *See also* Amalgamated Society of Engineers (ASE)

Amalgamated Society of Engineers (ASE), 95, 107, 114, 160–61n69. *See also* Amalgamated Engineering Union (AEU)

Amalgamated Workers, Brickmakers and General Laborers, 155n125

Anglo American Tin Stamping Co., 144n8

Anglo Enamelware, 70, 159n29

annealing, 20–21, 60

Anti-Suffragists, 153n110

Antonio, Robert J., 166n2

apprenticeship: in buckle-making venture, 13; conflict over, 42–45, 54–55; decline of, 63, 144n10; in organization of work, 16, 19, 26, 66, 114; terms of, 23, 38, 39, 40, 137n35, 140n33; in turner's termination agreement, 50. *See also* subcontracting; turners

arbitration. *See* collective bargaining

Archibald Kenrick and Sons: becomes a "controlled establishment," 101; Centennial celebration of, 46–49, 137n41, 141n50, 142n68; failure to post lockout notice, 114; founding of firm, 1; incorporation, 37; munitions produced by, 159n29; rules to be observed at, 67, 148n53. *See also* Ministry of Munitions (MUN)

Armstrong-Whitworth, 105

arson, 93–94, 97, 101, 102, 104, 156n3. *See also* class

artisan: in British history, 4–5, 55; decline of in metal trades, 63, 124, 131n20, 138n44; families, 62; at Kenricks, 14, 21, 26, 28, 38, 39, 56, 144n10; replaced by unskilled, 59. *See also* apprenticeship; deskilling

ASE. *See* Amalgamated Society of Engineers (ASE)

About the Authors

William G. Staples is professor of sociology at the University of Kansas. He has interests in social control, work, politics, and historical sociology. He is the author of *Castles of Our Conscience: Social Control and the American State, 1800-1985* (1991), *Culture of Surveillance* (1997), and *Everyday Surveillance: Vigilance and Visibility in Postmodern Life* (2000).

Clifford L. Staples is professor of sociology at the University of North Dakota. He has interests in social inequality, work, and the history of capitalism. His recent publications have appeared in the *Journal of Historical Sociology*, *Social Thought and Research*, and the *Review of Radical Political Economics*.